Digital Electronics:
A Primer
Introductory Logic Circuit Design

ICP Primers in Electronics and Computer Science

ISSN: 2054-4537

Series Editor: Mark Nixon *(University of Southampton, UK)*

Vol. 1 Digital Electronics: A Primer
 Introductory Logic Circuit Design
 by Mark Nixon

ICP Primers in Electronics and Computer Science Vol. 1

Digital Electronics:
A Primer
Introductory Logic Circuit Design

Mark Nixon
University of Southampton, UK

Imperial College Press

ICP

Published by

Imperial College Press
57 Shelton Street
Covent Garden
London WC2H 9HE

Distributed by

World Scientific Publishing Co. Pte. Ltd.
5 Toh Tuck Link, Singapore 596224
USA office: 27 Warren Street, Suite 401-402, Hackensack, NJ 07601
UK office: 57 Shelton Street, Covent Garden, London WC2H 9HE

Library of Congress Cataloging-in-Publication Data
Nixon, Mark (Mark Stewart)
　　Digital electronics : a primer - introductory logic circuit design / Mark Nixon, University of Southampton, UK.
　　　　pages cm. -- (ICP primers in electronics and computer science ; volume 1)
　　Includes bibliographical references and index.
　　ISBN 978-1-78326-489-6 (hardcover : alk. paper) -- ISBN 978-1-78326-490-2 (pbk. : alk. paper)
　　1. Digital electronics. 2. Electronic circuit design. I. Title.
　　TK7868.D5N52　2015
　　621.3815--dc23
　　　　　　　　　　　　　　　2014047465

British Library Cataloguing-in-Publication Data
A catalogue record for this book is available from the British Library.

Typeset by Stallion Press
Email: enquiries@stallionpress.com

Printed in Singapore

Preface

D igital electronics are pretty well everywhere these days: in car engines, doorbells, and the washing machine — it is basic, basic technology. The basics were established in the late 19th and early 20th centuries. Later in the 20th century there was an explosion of electronics, computers, and memory, leading to smaller, faster devices and cheaper large-scale memory. The technology has changed a lot too. Some early devices were so revolutionary that they caused textbooks to be centred on them. Other technologies were so influential that they still permeate textbooks, long after their obsolescence. This leads to a stock of textbooks with a lot of history in them that diverts from appreciation of the basic material.

This book takes a new approach, combining a textbook with working demonstrations. It centres on the essentials and is thus much more compact. It has developed from my lecturing on digital electronics over the past 30 years at the top department of electronics in the UK (well, it has been top of the league tables for all those years at least). In our courses we cover the basics of technologies, aiming for modern implementation. That is what this book is about.

In terms of style, example questions come at the end of each chapter. There are abbreviated solutions to the questions at the end of the book. There are quite a few logic simulators around these days, and there is support material for these within the book. There are worked examples for every circuit mentioned in this book using the Logisim simulator — these can help greatly the appreciation of basic issues. You will find these at the

book's website (www.users.ecs.soton.ac.uk/msn/digitsbook/), where you will also find other support material (e.g. for lecturers).

There has been a lot of input to this book and it is hard to acknowledge it all. I'm very grateful to the students I have lectured and who have pointed out corrections and extensions to the material herein. There has also been much advice from the academic staff at the School of Electronics and Computer Science at the University of Southampton, and I am much indebted to Adrian Pickering for his help. I am thankful to the excellent team I have been working with at Imperial College Press. My research students have played their part, and my family provide a great background upon which to do my work.

The production of a book has many stages, all of which can lead to error. Those errors are naturally my responsibility and so I am pleased to offer pints of fine English real ale for any errata that you may find herein (which haven't been found already — I'm not that generous!). The current version of the erratum sheet is to be found at the book's website.

I certainly hope you enjoy the book, and that it gives you a pleasant and digestible introduction to one of the basic technologies of our modern society.

Professor Mark Nixon

School of Electronics and Computer Science, University of Southampton

May 2014

Contents

Chapter 1

Introduction

1.1 Introducing Digital Electronics

Computers are at the heart of our daily existence, as are mobile phones. Both implement *software* using electronic circuits called *hardware*. The hardware is made up from *digital electronics*: these are circuits that can implement the software instructions. This book describes how you design the hardware, the digital electronic circuits. These circuits are used more widely than just in computers and phones. Around the home you will find them in many domestic appliances, even in front door chimes and kids' toys — researchers are even putting them into cloth to make it 'intelligent'.

This book starts with the basic devices and techniques and by its end you will have a good stock of building blocks and design techniques. The current position of digital electronics is a long way from its origins since it has progressed rapidly with the awesome developments in electronic (and particularly microelectronic) technology. Its origins lie in studies by mathematicians in the 19th century. One name might be familiar — although Lewis Carroll is perhaps most famous for *Alice in Wonderland* he was also, as Charles Dodgson, an eminent mathematician (Dodgson, 1958 [1887]). However, it was George Boole who published *An Investigation into the Laws of Thought* (1958 [1854]), which was such a major contribution that the algebra is called Boolean logic. This algebra had to await application in an engineering sense until Claude Shannon implemented it in switching circuits using relays, which were the earliest form of switching devices. Relay circuits progressed to valve circuits but the computers that were built using them were primitive, bulky and power-greedy. The next major

1

breakthrough was the semiconductor transistor, which miniaturised switching circuit implementation. When small numbers of switching circuits were combined (or integrated) in a single chip the technology was known as *small-scale integration* (SSI). As miniaturisation increased so did the number of devices on a chip, and microelectronic technology matured through *medium-scale integration* (MSI) to *large-scale integration* (LSI) in the 1970s to *very-large-scale integration* (VLSI) in the early 1980s. The term VLSI still applies, although *ultra-large-scale integration* (ULSI) has found some favour. Miniaturisation has continued and *microelectronics* seems rather dated now. The continuing progress is reflected in the newer term *nanoelectronics*. We now get several computers (cores) on a chip — and this was once a dream.

Design techniques have developed concurrently with these hardware developments and have even been motivated by technological development. Some older design techniques have little relevance now, having been developed concurrently with, say, the devices in MSI or LSI technology. The main developments in recent years have included programmable systems that require software specification of a device's function. Principles do not change, and that is what this book concentrates on (and why it was written!)

1.2 Organisation of This Book

The book starts with a study of combinational logic in Chapters 2 and 3. Chapter 2 covers the basics of combinational logic (how basic logic circuits can be designed to implement digital functions), whilst Chapter 3 covers how we reduce or minimise the circuits to achieve practical designs. This is the basis of the subject and several presentations of this can be found in the selection of texts in the reference section. The major difference in this book is that we shall concentrate on principles in a practical framework and in an approachable way.

Logic circuits are implemented using electronic circuits, which are described in Chapter 4. Computer science students can study these circuits to appreciate the advantages of particular technologies since these confer advantages to the computer systems that they program and interface. Electronic and electrical engineering students should note that

the treatment of microelectronic technology here concerns switching circuits only, although these are just branches of a much wider subject. Digital circuits are implemented using logic technologies that continue to evolve, and Chapter 4 is a snapshot (at the time of writing) of the current predominant technologies — this is CMOS (complementary metal–oxide–semiconductor), since it can combine high-speed operation with low power consumption.

A major part of digital electronics is sequential system design, which concerns digital circuits that go through a sequence of specified conditions, or states, as in a computer program. The main components are called 'bistables' and are described in Chapter 5. The main sequential design techniques are the focus of Chapter 6. The main design techniques here use positive edge-triggered bistables, and design systems using the algorithmic state machine (ASM) method.

A number of textbooks on digital electronics start with number systems and their arithmetic, concentrating on the binary system in particular, to introduce digital electronics as a vehicle for implementing binary arithmetic. This subject is covered in Chapter 7, in part to allow for inclusion of design examples concentrating on circuit design. This allows computer scientists in particular to appreciate exactly how the circuits at the heart of a computer are designed. The design examples concentrate on comparing the advantages of different approaches, contrasting results achieved with those of commercial designs.

There are design examples throughout and these predominate in the design chapters (3, 5, and 6). At the end of every chapter there are example questions, and abbreviated solutions to the questions are found in Appendix 1. There is other support material and most of the circuits used herein can be found in working form for a freely available logic simulator, and downloaded from the web (see the Preface for details). In this way, we aim to ensure you have a full preparation of the principles of logic design. The references made in the text are to be found at the end of the book. The earlier version of this text (Nixon, 1995) covered more ground. Here we concentrate on principles of logic circuit design and use a logic simulator (to make it more digestible), and we concentrate on principles to enable a tight focus on the technology that is central to our daily lives.

1.3 Preliminaries

1.3.1 *Binary numbers*

A number of assumptions have been made concerning prior knowledge and the most important of these are reviewed briefly here. Although binary arithmetic is studied in Chapter 7, earlier chapters do assume that you know what binary numbers are. The representation of a binary *integer* (without a fractional part) uses symbols 0 and 1 as *binary digits* or *bits* in the number. This can be compared with decimal arithmetic which has ten symbols going from 0 to 9, which are used for the *decimal digits*. In decimal arithmetic the four-digit integer 6203 is actually

$$6 \times 1000 + 2 \times 100 + 0 \times 10 + 3 = 6 \times 10^3 + 2 \times 10^2 + 0 \times 10^1 + 3 \times 10^0$$

Each decimal digit is then multiplied by a factor of ten. This factor corresponds to the position or significance of the digit, which increases from the right (in this case 3 is the least significant digit) to the left (and 6 is the most significant digit). The basis of the decimal system is 10 and that is why we have ten symbols and a factor of ten multiplying each.

The bit is the basic unit in digital electronics and in digital communications. It is derived from: *binary* + *digit* = *bit*. The basis of the binary system is 2: a bit can have one of two values and we have two symbols 0 and 1. The binary number 1100 (12 in decimal) is made up from bits as

$$1 \times 8 + 1 \times 4 + 0 \times 2 + 0 = 1 \times 2^3 + 1 \times 2^2 + 0 \times 2^1 + 0 \times 2^0$$

Since the base is smaller, and we have fewer symbols, the numbers need more digits. These are the bits and the bit with the highest factor of 2 (here 2^3) is the *most significant bit* (MSB) and the one that multiplies 2^0 (1) is the *least significant bit* (LSB). A four-bit binary count sequence compared with its decimal equivalent is given in Table 1.1.

The main advantage of the binary system is that it is very convenient for digital circuits in general and computers in particular. It is covered in much more detail later in Chapter 7, where other number systems are introduced, and we consider conversion between systems and circuits that implement computer arithmetic.

Table 1.1 Binary numbers

Decimal	Binary	Decimal	Binary	Decimal	Binary	Decimal	Binary
0	0 0 0 0	4	0 1 0 0	8	1 0 0 0	12	1 1 0 0
1	0 0 0 1	5	0 1 0 1	9	1 0 0 1	13	1 1 0 1
2	0 0 1 0	6	0 1 1 0	10	1 0 1 0	14	1 1 1 0
3	0 0 1 1	7	0 1 1 1	11	1 0 1 1	15	1 1 1 1

1.3.2 *(Very) basic circuit theory*

Electronic circuit theory is the mathematical theory describing the operation of electronic circuits. Here is a very basic introduction: the theory concerns in particular the relation between voltage (or potential) and current (the flow of charge or electrons). By analogy to water, voltage can be considered as water pressure and current as the rate of water flow. Water flows from the supply to the destination, induced by the pressure difference between the two. Similarly, current in electronic circuits is related to voltage. As in Figure 1.1, electronic current, denoted i, is shown by an arrow to flow from the positive power supply (+Volts) to the negative power supply (−Volts [usually ground or 0 volts]) and is depicted as opposite to the flow of electrons (which are negative charges, and thus attracted to the positive voltage supply).

Figure 1.1 Circuit characteristics

The relationship between voltage, resistance and current is given by Ohm's law as:

$$\text{Voltage } (V) = \text{Current } (i) \times \text{Resistance } (R)$$

This implies that if we increase the voltage for a constant resistance then the current increases as well, so by analogy greater water pressure increases the rate of flow. Also, for a constant voltage supply an increase in resistance implies a reduction in current. That is actually all we need to know about circuit theory for most of this book (in that we will not delve any deeper). Electronics and electrical engineering students will meet this as *analog circuit design*, which underlies the digital circuit design covered here.

1.3.3 *Circuit drawing*

There are several conventions for drawing logic circuits that are in common practice and will be used throughout this book. The first concerns how joins are depicted between circuit conductors. The join between one conductor and another, where the first one terminates, is shown in Figure 1.2(a). When the join is between two conductors where both continue is shown in Figure 1.2(b). The circuit where two conductors cross without intersection is shown in Figure 1.2(c) — there is no join between the two conductors in this circuit. (At the time of writing, the standard for 'structuring of the information about systems' is the International Standard IEC 81346-1, and this is a highly abbreviated form.) Note that if you follow the convention of Figures 1.2(a) and (b) precisely, then there is actually no need for the dot.

(a) Join (b) Join between two (c) No connection
 conductors

Figure 1.2 Drawing circuit connections

In practice, when conductors are joined together the circuit can sometimes be *incorrectly* depicted, as shown in Figure 1.3(a). This is avoided since it could be wrongly interpreted when, say, the circuit was built from a photocopied diagram of a circuit like Figure 1.2(c) and there was a small speck of dirt on the photocopier that made the circuit appear like Figure 1.3(a). An old-fashioned way of drawing the circuit in Figure 1.2(c) is shown in Figure 1.3(b) where little 'bridges' signify the absence of connection. This is *very* old-fashioned: imagine doing the little bumps

when one line is crossed by 64 others — that's a big ouch and we just don't do it that way these days. Don't draw circuits like Figure 1.3!

(a) Incorrect circuit connection (given potential confusion with Figure 1.2[c])

(b) Old-fashioned depiction of circuits crossing

Figure 1.3 Bad drawing styles

Another convention is in terms of direction of flow: inputs to circuits appear on the left and outputs appear on the right, as shown in Figure 1.4(a), and (when in circuits) do not have arrows indicating direction. In digital electronics, we can have connections made from multiple signals (on a 'bus'), shown in Figure 1.4(b) where each bus contains three signals (wires).

(a) Full circuit

(b) Using busses

Figure 1.4 Input/output convention

1.3.4 *Logic simulators*

This is where we are grateful to Carl Burch, who developed Logisim (http://ozark.hendrix.edu/~burch/logisim/) — a simple tool for simulating logic circuits. It is open-source, cross-platform (it runs on Linux, MacOS X, and Windows), and Java-based (Burch, n.d.):

> Logisim is an educational tool for designing and simulating digital logic circuits. With its simple toolbar interface and simulation of circuits as you build them, it is simple enough to facilitate learning the most basic concepts related to logic circuits. With the capacity to build larger circuits from smaller subcircuits, and to draw bundles of wires with a single mouse drag, Logisim can be used (and is used) to design and simulate entire CPUs for educational purposes.

Logisim is an excellent tool for introductory logic design. It is certainly not perfect as it misses out some points we shall need. Overall, though, it serves its task very well indeed and you can quickly draw up circuits and see how they operate. There are worksheets for each of the main logic design chapters (Chapters 2, 3, 5, and 6) and these can be found at www.users.ecs.soton.ac.uk/msn/digitsbook/. There is a new version of Logisim in development, but at the time this book was completed it was not available in operational form. There are others, such as the similarly named LogicSim (http://www.tetzl.de/java_logic_simulator.html), which largely serve the same purpose, but my students have found Logisim the easiest to install and use. Some of the disadvantages of Logisim are that it has no timing information and cannot draw waveforms, but these are easily outweighed by ease of use and clarity of display. Most of the documentation is on Logisim's website (as cited above) and there are links to two published papers by Burch (2002; 2004). I have used simulators in my lectures since their inception and my students find them extremely useful. Naturally, the simulators provide a way for you to check that your design really works, and that is always wonderful to see.

1.4 Concluding Comments and Further Reading

The preliminaries presented in this chapter serve as a basic introduction to the remainder of the book. We shall start with combinational circuit design — the beginning of digital circuit design and the basis of the whole subject. The circuit theory overviewed here will not be used until Chapter 4, but it's worth practising it in the following example questions. For those who would like more background on electronic circuit theory, have a look at Sedra and Smith (2010) or Spencer and Ghausi (2003) (and there are many other similar texts).

We will be using the above-discussed drawing conventions, binary numbers, and Logisim in the next chapter. You can either memorise Table 1.1, or go to Chapter 7 and have a read there. It is interesting stuff, but is placed later in the book so we can now head straight into the design of basic digital circuits.

1.5 Questions

Solutions in Appendix 1.

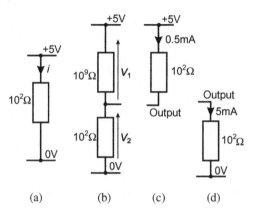

Figure 1.5 (Very) basic circuits

Q1.1 For the circuit in Figure 1.5(a), determine the current i.

Q1.2 For the circuit in Figure 1.5(b), determine the voltages across the two resistors.

Q1.3 For the circuit in Figures 1.5(c) and (d), determine the output voltage in each case.

> *Don't forget this analysis, as we are actually describing the outputs of logic gates and that will be covered in Chapter 4.*

Q1.4 Install Logisim (http://ozark.hendrix.edu/_burch/logisim).

Chapter 2

Combinational Logic Design

2.1 Combinational Logic/Boolean Algebra

This book is about introductory digital electronic design. The focus is on design and so we shall start with the building blocks we use to make up designs. As such, we shall leave the formal descriptions until later, and here we shall use *Boolean algebra* to develop digital electronics. In its broader form Boolean algebra is known as discrete mathematics, which, amongst other subjects, has wide application and implication in software engineering. This introduction is functional and serves as a basis for design. For deeper study (albeit sometimes pedagogic to an engineer) you can refer to one of the texts listed in Section 2.7.

One definition of *logic* is as a chain or science of reasoning. For Boolean algebra we are concerned only with truth or falsity. We are concerned with propositions that can have either of two values: FALSE or TRUE. A *proposition* is a variable or fact. For example, the proposition 'The contact lens is not cracked' is TRUE if the lens has no crack in it, but FALSE if the lens is actually cracked. There is no halfway house here: propositions are only TRUE or FALSE.

A *statement* is formed by joining propositions; the overall validity of the statement depends on the validity of each proposition from which it is formed. Consider, for example, the statement 'A good contact lens is not cracked and circular'. (NB. This will later be formulated in terms of industrial quality control). If the proposition 'The contact lens is not cracked' is TRUE and the proposition 'The contact lens is circular' is TRUE then the statement 'The contact lens is good' is TRUE, since it is neither cracked nor an ellipse. This can be summarised using a *truth table*, which

shows the validity of a statement according to the validity of its constituent propositions.

> *Note here that the contact lens can only be cracked or not — just one of two values. A contact lens is not slightly cracked or partly broken (if it was cracked you wouldn't buy it). This is in accordance with Boolean algebra where variables can have two states only — though in practice small cracks invisible to the human eye are doubtless fine (depending on who's wearing the lens).*

A truth table for a statement made from two propositions is a list of the validity according to the four possible sets of the two propositions. For propositions *A* and *B* the four sets or combinations are:

(i) both FALSE; (ii) *A* FALSE, *B* TRUE;
 (iii) *A* TRUE, *B* FALSE; and (iv) both TRUE.

Using FALSE = F and TRUE = T, the truth table for a statement which is a function of these two variables, $f(A, B)$, is:

A	B	$f(A, B)$
F	F	
F	T	
T	F	
T	T	

The question then remains to determine whether the output is TRUE or FALSE for each input combination. Take the statement 'A good contact lens is not cracked and circular'. This can be broken into a statement 'The contact lens is good' and two propositions 'The contact lens is not cracked' and 'The contact lens is circular'. The truth table then indicates when or not the contact lens is good according to whether it is cracked (or not) or circular (or not):

A = proposition 1, 'The contact lens is not cracked' (TRUE = unbroken, FALSE = broken)

B = proposition 2, 'The contact lens is circular' (TRUE = circular, FALSE = elliptical)

$f(A, B)$ = statement, 'The contact lens is good'

Propositions			Statement
A	B	$f(A, B)$	
F	F	F	The contact lens is cracked and not circular, it is not a good lens.
F	T	F	The contact lens is round but no good since it is cracked.
T	F	F	The contact lens is not cracked but it is not round: it's no good.
T	T	T	The contact lens is good since it is not cracked and it is circular.

2.2 Logic Functions

What we have actually defined is a way of linking variables to form a result. As in traditional mathematics, the variables are linked by a function. The function that we have just seen illustrated is actually the *AND* function, which is perhaps unsurprising since the propositions were connected using the word 'and'. The truth table for the AND function is then

A	B	A AND B
F	F	F
F	T	F
T	F	F
T	T	T

This shows that the AND function is identical to the way that we use it in language. The function A AND B is only TRUE when A is TRUE and B is TRUE, otherwise it is FALSE.

There is also a logical *OR* function which again follows the way we use it in language. Consider the following statement: 'I will watch television if I have nothing else to do or there is a programme I really want to see.' Given the statement 'I will watch television', this will be TRUE if either or both the propositions 'There is a programme I really want to see' and

'I have nothing else to do' are TRUE. The OR function can be summarised by its truth table

A	B	A OR B
F	F	F
F	T	T
T	F	T
T	T	T

As with the AND function, the word 'or' connects the two propositions in the original statement, so it is used in a logical way and in a similar way that we use it in language.

The AND and OR functions are basic logic functions. The last member of the basic set is called *NOT*. Its function is implicit in its name; it provides a logical **complement** or *inversion*. NOT TRUE is FALSE; NOT FALSE is TRUE, which gives a truth table

A	NOT A
F	T
T	F

These functions are actually mathematical functions and there is a **logic algebra** formulated around them. This is the basis of discrete mathematics. The aim here has been to introduce combinational logic as part of digital electronics; we shall now move to logic circuits, and then define the logic algebra later, in the context of combinational logic design.

2.3 Combinational Logic Implementation

We have so far considered Boolean algebra as a formalism to demonstrate logical relations. Consider, for example, the problem of the contact lens, which when on a production line has a natural system implementation: a quality control system to remove the lens from the production line.

In your studies away from digital electronics, you will doubtless learn how to design systems to measure whether an article is cracked (can we use ultrasound or computer vision?) or whether it is circular (how do we measure shape?), and systems to remove objects from a production line. Before that though, we need to design a system that can interpret the

Figure 2.1 On quality control

signals from the sensors, to combine them together to provide a signal that stimulates removal of the contact lens. Generally, that is part of a computer program. Here we are at the heart of the matter: how do we combine signals to achieve a desired function? This is combinational logic, the implementation of Boolean algebra in switching circuits. Logical variables, *inputs*, are connected by logical functions, *gates*, to give a logical result, an *output*. A gate is an electronic circuit that implements a logic function via a switching circuit.

2.3.1 *Combinational logic levels*

We need signals to represent TRUE and FALSE. These are necessarily electronic signals since we shall implement them in electronic circuits. We could use current or voltage and both are indeed used in logic circuits. First though we shall use *level logic* — voltage levels that denote TRUE and FALSE.

$$\text{TRUE} = \text{`1'} = +5\,\text{V} \quad \text{and} \quad \text{FALSE} = \text{`0'} = 0\,\text{V}$$

Modern logic systems use a voltage lower than $+5$ V, but this remains a good starting point. We use the quote marks to denote logical variables. This is because we are using a binary system with just two values, and this is just one way of indicating it.

The use of an explicit level for TRUE and FALSE would be very difficult to achieve. We actually use a range of levels (for reliability).

The use of a high level to indicate a '1' and a low level to represent a '0' is called *positive logic*. You will meet this often in *manufacturers' data* which define how circuits operate.

2.3.2 Combinational logic gates

2.3.2.1 AND

The AND gate (Figure 2.2) is a circuit implementing the logical AND function. The AND function is TRUE when both propositions are TRUE, the output of an AND gate is '1' when both inputs are '1'. If either input is a '0' then the output is '0'.

Truth Table AND Circuit Symbol

A	B	$A \cdot B$
0	0	0
0	1	0
1	0	0
1	1	1

Inputs Output

Figure 2.2 AND gate

The symbol · symbolises the AND function — it looks like multiplication and in a loose way we can now think of the AND function as multiplication (anything multiplied by zero results in zero). As in algebra, the · can be (and often is) omitted. In circuit diagrams we need a symbol for the AND function. The symbol for AND is like the letter D in AND, but stretched horizontally. In terms of a switching circuit we can visualise the AND function as the collection of electrical switches to switch on a light; we connect a power supply, V_{cc}, so that current flows through the switches illuminating the light to ground (0 V). The buttons are normally open (which represents '0') and close when pressed (representing '1'); the light being on represents a '1' (Figure 2.3).

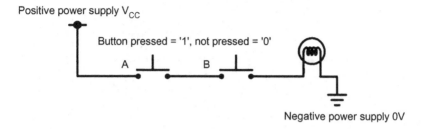

Figure 2.3 AND gate equivalent circuit

In the AND function we need both inputs to be a '1' before the output is a '1'; here we need both buttons to be pressed before the light is switched on. This can be seen in the operation of the gate: the lamp will be illuminated only when both inputs are at '1'. This is the same for an AND gate: both inputs at '1' and the output is '1' (and a bit lighter!); any input at '0' and the output is '0' (the output is black), as in Figure 2.4.

Figure 2.4 AND gate function

2.3.2.2 *OR*

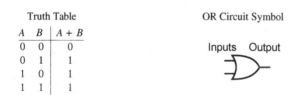

Truth Table			OR Circuit Symbol
A	B	A + B	
0	0	0	Inputs Output
0	1	1	
1	0	1	
1	1	1	

Figure 2.5 OR gate

The OR function (Figure 2.5) is symbolised by addition, which forms a useful memoriser. It is a loose mnemonic since $1 + 1 = 1$! (We shall study arithmetic systems later and find the proper circuit for addition). The switching circuit for OR is given by two switches in parallel. Should we press either or both buttons then the lamp will light (Figure 2.6).

Button pressed = '1', not pressed = '0'

A

B

Negative power supply 0V

Figure 2.6 OR gate equivalent circuit

Note that diodes can be used to replace the switches in the OR circuit. If the diodes' anodes are connected to logic inputs, and their cathodes are connected to the lamp then we effectively form a diode OR gate.

2.3.2.3 *NOT/Inverter*

The NOT function (Figure 2.7) is symbolised by placing a bar across a logic variable: \overline{A} represents NOT A. It has to be drawn carefully so as to cover all factors that are to be inverted.

Truth Table

A	\overline{A}
0	1
1	0

NOT/ Inverter Symbol

Input Output

Figure 2.7 NOT gate

2.3.3 *Circuits and truth tables for more than two inputs*

Truth tables and gates have so far been designed in terms of two inputs only. The truth table is merely a list showing the value of an output (or statement) for a combination of inputs (propositions). It can easily be extended to handle more than two input variables. Consider a function: 'A cup is good if it has a handle, it is unbroken and it is glazed.' This function is TRUE when all three propositions are TRUE. In terms of a truth table, we need to tabulate whether or not the output is '1' (TRUE) according to the status of the three propositions. Each of the propositions can have either of two values and so there are $2 \times 2 \times 2$ (i.e. 8) combinations. The truth table is

then formed from the variables or propositions. The propositions are:

$A =$ the cup has a handle ('1' = TRUE = it has a handle)
$B =$ the cup is unbroken ('1' = TRUE = unbroken)
$C =$ the cup is glazed ('1' = TRUE = glazed, '0' = unglazed)
$f(A, B, C) =$ the cup is good ('1' = TRUE = good cup)

The truth table, Figure 2.8, then covers the eight combinations of A, B, and C, giving the output value $f(A, B, C)$ for each combination. The function is TRUE only when all inputs are TRUE and therefore describes a three-input AND function. It is implemented using a three-input AND gate which provides, as output, a '1' (5 V) when all inputs are '1' (5 V) and its symbol is the AND symbol with three input connections. (For a switching circuit, we could simply add in a third switch for C in series with the ones for A and B to the circuit in Figure 2.3.)

There are two ways to construct a truth table. One is to list all input combinations by counting up in binary. The other way is to take an input and as a column write it down as 01010101, take the next column and write its values as 00110011, and the last as 00001111. Note that if you go too far (with too many rows), you'll just repeat the truth table (starting with a second entry for 000).

Truth Table				3-input AND Symbol

A	B	C	$f(A,\ B,\ C) = A \cdot B \cdot C$
0	0	0	0
0	0	1	0
0	1	0	0
0	1	1	0
1	0	0	0
1	0	1	0
1	1	0	0
1	1	1	1

Inputs Output

Figure 2.8 Three-input AND gate

There is also a three-input OR function which is TRUE if one or more inputs is TRUE. It is implemented using a three-input OR gate which provides an output '1' (5 V) when any of the inputs is '1' (5 V) (Figure 2.9).

Truth Table 3-input OR Symbol

A	B	C	$f(A, B, C) = A + B + C$
0	0	0	0
0	0	1	1
0	1	0	1
0	1	1	1
1	0	0	1
1	0	1	1
1	1	0	1
1	1	1	1

Inputs Output

Figure 2.9 Three-input OR gate

This can of course be extended into truth tables that describe circuits with four inputs or more. Note that for n inputs, there are 2^n combinations. A truth table for ten inputs tabulates the value of the output for 2^{10} (i.e. 1024) combinations and is rather large. Not all circuits reduce to a simple gate implementation such as a three-input AND gate, most usually reduce to a collection of different gates.

2.3.4 *Don't-care inputs*

When implementing designs there are sometimes input combinations where the status of one input cannot affect the output. This may be because either it has no effect in that combination, or because it physically cannot occur. These are called *don't-care inputs* and they are usually denoted using an X in the truth table; this X signifies that the signal can be '0' or '1' and its value has no consequence on the output in that state. For an input B whose value cannot change the output function f when the input A is '1', the following truth tables are equivalent:

A	B	$f(A, B)$
1	X	1

\rightarrow

A	B	$f(A, B)$
1	0	1
1	1	1

2.3.5 *Further logic gates*

Two other important gates are formed by combining NOT with AND and OR. The bubble on the inverter symbol actually symbolises inversion, so the combination of NOT and AND, called *NAND*, is given by the

inversion of AND (NOT AND) and is symbolised by the NOT bar across the AND function, and by the inversion bubble attached to the AND symbol (Figure 2.10).

Truth Table NAND Symbol

A	B	$A \cdot B$	$\overline{A \cdot B}$
0	0	0	1
0	1	0	1
1	0	0	1
1	1	1	0

Inputs Output

Figure 2.10 NAND gate

A switching circuit can again be constructed but in this case by using switches which are normally closed (representing '0' here) and open when pressed (representing a '1') (Figure 2.11).

Figure 2.11 NAND gate function

Only when both buttons are pressed does the lamp switch off. If one or neither button is pressed, the lamp lights.

The AND circuit is formed by a series of switches whereas for an OR function they are in parallel. For NAND the switches are in parallel and for NOR in series, but with normally closed switches rather than normally open switches. This illustrates that the NAND gate can be implemented as an OR gate with inverted inputs; it is often drawn this way, as part of a circuit drawing convention.

The combination of NOT and OR gives NOT (OR), called *NOR*. The truth table and symbol are as per Figure 2.12, with a switching circuit again using normally closed switches but this time connected in series.

		Truth Table		NOR Symbol

A	B	$A + B$	$\overline{A + B}$
0	0	0	1
0	1	1	0
1	0	1	0
1	1	1	0

Figure 2.12 NOR gate

Figure 2.13 NOR gate function

Here the lamp is switched off when either button is pressed (Figure 2.13).

2.3.6 *Other symbols*

Other logic symbols are used, for example those defined according to the British Standards Institution (BSI) (BS3939:1985; Figure 2.14).

Figure 2.14 BSI symbols

These symbols have found usage, but here we have used the symbols from standard IEEE/ANSI 91:1984, which derived from a US Military Standard set. These symbols will be used for logic gates throughout this

book. The bubble signifying inversion can actually be attached either to inputs or to outputs, giving two possible representations for an inverter.

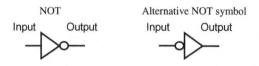

Figure 2.15 Mixed logic drawing

This is part of a drawing convention known as *mixed logic*, a convention for drawing circuit diagrams (which we shall avoid explaining as it's very confusing for an introductory view) (Figure 2.15).

2.3.7 *Complete set of two-input logic functions*

For a two-input logic gate there are four possible input combinations. For each of the combinations there are two possible output values. A two-input logic gate then has $2^4 = 16$ possible output sets:

Inputs		Possible Output Sets															
A	B	0	1	2	3	4	5	6	7	8	9	10	11	12	13	14	15
0	0	0	0	0	0	0	0	0	0	1	1	1	1	1	1	1	1
0	1	0	0	0	0	1	1	1	1	0	0	0	0	1	1	1	1
1	0	0	0	1	1	0	0	1	1	0	0	1	1	0	0	1	1
1	1	0	1	0	1	0	1	0	1	0	1	0	1	0	1	0	1

By inspection we can recognise some of these. Set 1 represents AND, set 7 is the OR function, set 8 is NOR, and set 14 is NAND. Some are evidently recognisable and equally evidently not a logic gate: set 0 is '0' for all combinations, whilst set 15 is '1' for all combinations. As logic functions they are called FALSE and TRUE, respectively; there is no gate which implements them since it wouldn't be any use (if we want a '0' or a '1' we use the power supply since that's what it's there for). Set 3 is a direct copy of A, set 5 is B, set 10 is \overline{B}, and set 12 is \overline{A}. The existence of these as a function is a consequence of determining all possible output combinations. Others are equally simple: Set 2 is '1' in one case, for $A =$ '1' and $B =$ '0'. This is then $A \cdot \overline{B}$ and so set 2 represents an AND gate with one input inverted. So does set 4, $\overline{A} \cdot B$. Set 11 is $A + \overline{B}$ and set 13 is $\overline{A} + B$; again

there is no gate to implement them. Only two combinations remain, set 6 and set 9, and for combinational logic these are functions in their own right. Set 6 represents the OR but excludes the case when the inputs are both '1'; in consequence it is called *exclusive OR* or *XOR* (Figure 2.16). Set 9 is the logical inversion of set 6 and is called *exclusive NOR* or *XNOR* (Figure 2.16). The XNOR function is '1' when the inputs are the same is for this reason is often called *equivalence*. All functions in this table are actually named in discrete mathematics, but except for AND, NAND, OR, NOR, or NOT, only XNOR and XOR have any consequence in combinational logic design. The final member of the set of logic functions is a *buffer*, whose output follows the input (Figure 2.16).

The functions that are gates are actually those which are symmetric in both inputs. That way it does not matter which way round they are connected. We shall come to this later, when we consider logic laws.

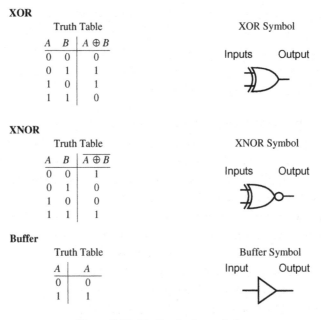

Figure 2.16 Further logic symbols

A buffer is used to distribute and *interface* (connect) signals in logic circuits. Its symbol is that of an amplifier and that is its function: it is used to maintain power (or to top it up) in practical circuits. Logic gates can only source limited amounts of current and thus only drive a limited number of gates. If we need to distribute a signal to more than its specified maximum via a particular gate, then we need to buffer the signal and feed it via the buffer circuits. This coincidentally explains the use of the bubble to signify inversion — by its omission, the buffer is not NOT! These complete the full set of available functions.

2.4 Integrated Circuits

AND, OR, and NOT gates are available as *integrated circuits* or *chips* that give packages of gates. The circuit is actually inside the chip and the legs provide connections to your circuit (Figure 2.17).

Figure 2.17 Dual in line (DIL) chip

There is an alignment marker to ensure that you use a chip the right way round. There are two pins for the *power supplies* and the remainder are for connections to the circuits inside. There are many varieties of chips; common circuits are usually available in a *logic family* which is a series of common logic functions implemented using a particular *logic technology*. The most famous and enduring logic family was the 74 series introduced by Texas Instruments. Though it is largely obsolescent now, it remains an excellent introduction to logic technology. Amongst its members was the 7404 which provided six NOT gates (hex NOT). The six NOT gates

together with two power supplies (note the 5 V power supply — modern systems use a lower voltage) result in a 14-pin chip arranged as shown in Figure 2.18. The other power supply is ground (GND) which is the negative power supply 0 V.

The 74 series was particularly enduring in that it was implemented in differing logic technologies (Transistor–Transistor Logic [TTL] and CMOS). Any logic technology is a compromise between many factors, of which speed and power are amongst the most important, as considered in Chapter 4.

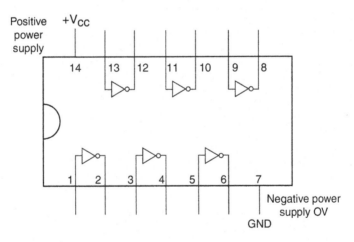

Figure 2.18 7404 NOT DIL chip

The 74 series included many more chips, which extended to combinations of gates. Less use is made of these chips now as *programmable logic* dominates implementation. Programmable logic chips offer a set of gates and the interconnection between these gates can be specified by software, thus controlling the chip's function. The chip then has a clearly documented design and reprogrammable chips can easily be reconfigured. These chips are a bit complex for this stage of the book.

2.5 Basic Logic Devices

2.5.1 *Coders and decoders*

The function of a *decoder* is naturally to **decipher** encoded information. Information is encoded primarily to maximise usage of channel capacity, such as for satellite communication where the available power is limited. By way of example, consider a traffic-light control system where a controller determines the sequence and timing of the lights, but then needs to turn them on and off (Figure 2.19).

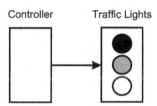

Figure 2.19 Traffic light control

The controller could provide three signals, one to turn on each light. To save money we could use just two wires and transmit logic signals by noting that two logic signals have four possible combinations. We can then design a *coder* which encodes the chosen information. By choosing the signals as:

Transmitted signals		A	B
illuminate red bulb	R	0	0
illuminate yellow bulb	Y	0	1
illuminate green bulb	G	1	0

In software this would be of the form:

```
IF red THEN BEGIN
            A=LOW
            B=LOW
            END                This software could also have been
IF yellow THEN BEGIN           written using a CASE statement which
            A=LOW              would have simplified it slightly.
            B=HIGH
            END
```

```
IF green THEN BEGIN
         A=HIGH
         B=LOW
         END
```

We can then use two lines to communicate between the controller and the lights. The signals need to be decoded from the chosen coding scheme to light the correct bulb. We then need a *decoder*. In this case, the specification for the decoder is:

Received Signals		Illuminated Light	
A	B		
0	0	R	illuminate red bulb
0	1	Y	yellow
1	0	G	green
1	1	U	unused combination

This is by way of introduction, since the decoder is actually a basic building block in digital electronics and there are standard circuits.

2.5.2 *2-4 line decoder*

A decoder that decodes the four possible combinations of two input signals is known as a *2-4 line decoder*. Its function is given by the truth table in Figure 2.20. This shows which input combination selects which channel.

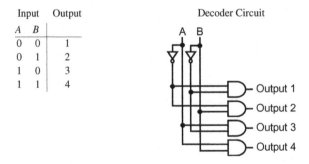

Input		Output
A	*B*	
0	0	1
0	1	2
1	0	3
1	1	4

Figure 2.20 Decoder

The logic circuit that implements the 2-4 line decoder is shown in Figure 2.20. If both inputs *A* and *B* are '0' then Output 1 is selected and

is '1' and all the other outputs are '0'. If *A* is '0' and *B* is '1' then the second output is selected and the others are unselected. The four cases of the circuit's operation are the four conditions for which the output channels are selected are shown in Figure 2.21 where a '1' is depicted in light grey and '0' in black.

Figure 2.21 The four states of a decoder

The *decoder symbol* is a rectangle, shown in Figure 2.22. In this symbol, the numbering of the output channels is from 1 to 4, so channel 1 is Output 1. The input selection is by two bits and these are shown underneath the decoder symbol. There are other symbols for a decoder, and this is the most common.

Figure 2.22 Decoder symbol

A software description of the 2-4 line decoder is:

```
CASE (A,B) ;comments follow the semicolon and are not
            executed
BEGIN
   00: BEGIN channel_0=1 ;If A=0 AND B=0 THEN channel_0=1
        channel_1=0, channel_2=0, channel_3=0
        ; The others are 0
      END
   01: BEGIN channel_1=1 ;If A=0 AND B=1 THEN channel_1=1
        channel_0=0, channel_2=0, channel_3=0
        ; The others are 0
      END
```

```
 10: BEGIN channel_2=1 ;If A=1 AND B=0 THEN channel_2=1
         channel_0=0, channel_1=0, channel_3=0
         ; The others are 0
     END
 11: BEGIN channel_3=1 ;If A=1 AND B=1 THEN channel_3=1
         channel_0=0, channel_1=0, channel_2=0
         ; The others are 0
     END
END
```

2.5.3 *Multiplexers and demultiplexers*

The function of a *multiplexer* is to choose an output from a selection of inputs in a manner similar to a rotary switch, which for two inputs is shown in Figure 2.23.

Figure 2.23 Multiplexer function

A *multiplexer symbol* is shown in Figure 2.24.

Figure 2.24 Multiplexer symbol

The Select input determines which of these two buttons is connected to the circuit's output.

Its function is described in software as

```
IF (channel_select) THEN output = Channel_2
      ;control = 1 connects Channel_2 to the output
            ELSE output = Channel_1
      ;control = 0 connects Channel_1 to the output
```

The function of a *demultiplexer* is the converse, to choose an output signal. In software its function is:

```
IF (control) THEN output_2 = input
      ;control = 1 connects output_2 to the input
         ELSE output_1 = input
      ;control = 0 connects output_1 to the input
```

There are other forms of multiplexing such as time-domain multiplexing where signals occupy a communication channel for specified time slots (e.g. in radio communications). The function of the multiplexer is then to choose from the input signals at the appropriate time slot.

2.5.4 4-1 line multiplexer

A circuit which chooses a single output from one of four inputs is called a *4-1 line multiplexer* (Figure 2.25). Note that we need to choose which of four lines is to be selected and then provided as the circuit output. Two logic signals provide four combinations and the channel to be chosen is then encoded using two select lines.

Select Inputs		Output
A	*B*	
0	0	Input 1
0	1	Input 2
1	0	Input 3
1	1	Input 4

4-1 Line Multiplexer Symbol

Figure 2.25 4-1 line multiplexer

Here the input selection is by two bits and this is shown underneath the decoder symbol. The circuit for a 4-1 line multiplexer, shown in Figure 2.26, then gates the channels with the select lines. The output is an OR gate as this combines the selected channel with signals which are all at '0' from the channels which are not selected (the output of one AND gate follows

a selected channel whereas the output of all the other AND gates is '0'). In this way, only the selected channel can change the output of the circuit to '1' since all the other inputs to the OR gate are '0'.

Figure 2.26 4-1 line multiplexer circuit

A software description is:

```
CASE (A,B)
BEGIN
  00: output=channel_1 ;If A=0 AND B=0 THEN output=channel_1
  01: output=channel_2 ;If A=0 AND B=1 THEN output=channel_2
  10: output=channel_3 ;If A=1 AND B=0 THEN output=channel_3
  11: output=channel_4 ;If A=1 AND B=1 THEN output=channel_4
END
```

Note that the specification on channel selection is identical to that for the 1-4 line decoder. This is hardly surprising since we are decoding two input lines to determine which of four inputs should be selected as output. The decoder signals are enabled when a particular channel is chosen and the multiplexer output is then given by ANDing the decoder outputs with the appropriate input signal. The outputs of the AND gates are then combined with a four-input OR gate whose output will follow the selected channel. Figure 2.27 shows a working multiplexer, from a Logisim implementation. The circuit is the same as that for Figure 2.26 with two small changes. First, the input channels have signal generators attached and second the OR gate, which collects the gated inputs, is a five-input gate (with one input set to '0') since Logisim does not allow four-input gates. The lighter connections reflect a signal at '1' and the darker lines a '0'. In Figure 2.27(a)

then connects Channel 1 to the Output when the select inputs *A* and *B* are both '0'. When A = '0' and B = '1' the output is derived from Channel 2, as in Figure 2.27(b). The same channel is connected in Figure 2.27(c) but the signal is '0' not '1' (note that Channels 3 and 4 are both '1' but are ignored).

(a) Channel 1 connected (b) Channel 2 (at '1')

(c) Channel 2 (at '0')

Figure 2.27 Multiplexer operation

The function is then defined by a truth table using don't-care inputs for the channels that are not selected. The output has the same (logic) value as the selected input channel.

| Input Channel Status | | | | Select Lines | | Output Channel | |
0	1	2	3	A	B	Selected	
0	X	X	X	0	0	0	}1
1	X	X	X	0	0	1	
X	0	X	X	0	1	0	}2
X	1	X	X	0	1	1	
X	X	0	X	1	0	0	}3
X	X	1	X	1	0	1	
X	X	X	0	1	1	0	}4
X	X	X	1	1	1	1	

2.6 Designing a Combinational Logic Circuit

The design of a circuit can be extracted from a truth-table specification. This is to be expected since a truth table is a complete specification of a circuit's operation. In a basic form the design can be implemented using AND, OR, or NOT gates.

The basis for such design is that the function will be '1' for various combinations of inputs. These input combinations can be grouped using AND gates, since the output of an AND gate is '1' when all its inputs are a '1'. The output of the circuit should be '1' if any of these combinations produce a '1', and we can therefore group the AND gate outputs using an OR gate, since the output of an OR gate is '1' if any of its inputs are '1'. This procedure is reflected in the earlier circuit for the multiplexer where the gated decoder outputs were collected using an OR gate. The procedure for design is then:

1. Fill in the truth table.
2. Extract conditions for which the output = '1'.
3. Group the conditions to form the final output.

The process is best illustrated by example. Consider the following 'specification'.

John Wonderland is going on holiday with his wife Alice, his beautiful daughter Bo and her friends, Chas and Dave. He wants to fish peacefully while his wife chaperones Bo, so he decides to make an alarm that goes off

when Bo is with either Chas or Dave (or both) but not Alice, either inside or outside the house. He buys radio transmitters to indicate whether or not people are inside the house and surreptitiously fixes them to everyone. He then needs an alarm circuit to act on the information from the transmitted signals.

This is clearly a fictitious problem. My students have suggested that they would steal Alice's transmitter and fix it to Bo.

We shall follow the stated design procedure and the first step is to draw up the truth table.

We shall use $A =$ Alice, $B =$ Bo, $C =$ Chas,
$D =$ Dave and $f =$ alarm;
and for all inputs, '1' = inside house, '0' = outside house;
and for the alarm '1' = alarm rings, '0' = alarm is silent.

The truth table is then:

A	B	C	D	f		
0	0	0	0	0		All outside together
0	0	0	1	0		Alice and Bo outside together
0	0	1	0	0		Alice and Bo together
0	0	1	1	0		Alice and Bo together
0	1	0	0	0		Bo inside alone
0	1	0	1	1	f_1	Bo inside with Dave
0	1	1	0	1	f_2	Bo inside with Chas
0	1	1	1	1	f_3	Bo with Chas and Dave
1	0	0	0	1	f_4	Alice inside, Bo outside with Chas and Dave
1	0	0	1	1	f_5	Alice inside, Bo outside with Chas only
1	0	1	0	1	f_6	Alice inside, Bo outside with Dave
1	0	1	1	0		Bo outside alone
1	1	0	0	0		Bo and Alice inside
1	1	0	1	0		Alice, Bo and Dave inside
1	1	1	0	0		Alice, Bo and Chas inside
1	1	1	1	0		All inside

The output is true in six of the 16 cases; in all others Bo is chaperoned by Alice, or alone, and the alarm does not go off. These six cases are labelled f_1, f_2, f_3, f_4, f_5, and f_6. In each of these cases a combination of inputs forces the output to be '1'.

For f_1 the output $=$ '1' for $A =$ '0', $B =$ '1', $C =$ '0', and $D =$ '1'
i.e. when $\overline{A} =$ '1', $B =$ '1', $\overline{C} =$ '1', and $D =$ '1'

f_1 is now expressed as variables which are all '1' which can thus be grouped using a four-input AND gate. f_1 can now be implemented as

$$f_1 = \overline{A} \cdot B \cdot \overline{C} \cdot D$$

and the output will be '1' when Alice and Chas are outside and Bo and Dave are inside. Note that f_1 will be '0' for all other input combinations.

We can do the same for f_2, so $f_2 =$ '1' for $A =$ '0', $B =$ '1',
$C =$ '1', and $D =$ '0'
i.e. when $\overline{A} =$ '1', $B =$ '1', $C =$ '1', and $\overline{D} =$ '1'

which can be implemented as

$$f_2 = \overline{A} \cdot B \cdot C \cdot \overline{D}$$

which is '1' only when Alice and Dave are outside, and Bo is inside with Chas.

In the same way, $f_3 = \overline{A} \cdot B \cdot C \cdot D$, $f_4 = A \cdot \overline{B} \cdot \overline{C} \cdot \overline{D}$, $f_5 = A \cdot \overline{B} \cdot \overline{C} \cdot D$, and $f_6 = A \cdot \overline{B} \cdot C \cdot \overline{D}$.

We now have six circuits that provide the alarm if their particular input combination is satisfied. We want to collect together the six terms to provide a single alarm. We need a circuit whose output is '1' when any of the inputs are a '1' — this is an OR gate. We can then collect f_1, f_2, f_3, f_4, f_5, and f_6 together, using

This then gives the full alarm circuit to provide the specified alarm according to the location of the conspirators which is:

$$alarm = \overline{A} \cdot B \cdot \overline{C} \cdot D + \overline{A} \cdot B \cdot C \cdot \overline{D} + \overline{A} \cdot B \cdot C \cdot D$$
$$+ A \cdot \overline{B} \cdot \overline{C} \cdot \overline{D} + A \cdot \overline{B} \cdot \overline{C} \cdot D + A \cdot \overline{B} \cdot C \cdot \overline{D}$$

The circuit for the alarm is given in Figure 2.28.

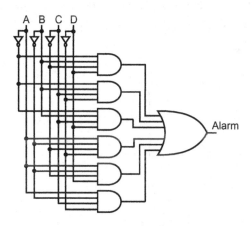

Figure 2.28 Wonderland alarm

This circuit uses more gates than it needs to and it can be reduced or minimised (see Section 3.3).

A software description for the alarm function for the same inputs is:

```
IF ((NOT(alice) AND bo AND chas AND dave)
OR (NOT(alice) AND bo AND chas AND NOT(dave))
OR (NOT(alice) AND bo AND NOT(chas) AND dave)
OR (alice AND NOT(bo) AND NOT(chas) AND NOT(dave))
```

```
OR (alice AND NOT(bo) AND NOT(chas) AND dave)
OR (alice AND NOT(bo) AND chas AND NOT(dave)))
   THEN alarm=TRUE
   ELSE alarm=FALSE
```

This software uses more words than it needs to (it is too prolix) and it can be reduced or minimise (see Section 3.3).

We have actually decoded the six cases where we want the output alarm to ring. As such, we could use a decoder (Figure 2.29). The alarm can ring in any one of the six cases, so we group the outputs using an OR gate. There are wasted gates since those that decode the unused input combinations lead to outputs that never change and do not contribute to the alarm function. Note that the unused decoder outputs are left unconnected, since they do not affect the alarm.

Figure 2.29 Wonderland alarm via decoder

We could also use a multiplexer, since that is largely a decoder controlling inputs to an OR gate. To do this we can use logic levels as the input channels and connect the inputs *ABCD* to the select inputs. We connect a '0' to the channels for which the alarm should not ring, and '1' to channels where the alarm will ring. Then, since the function of the multiplexer is to select from the inputs according to the values of the signals on the select lines, the input '1's to the multiplexer will be selected by those (six) combinations of *ABCD* for which the alarm should ring.

Figure 2.30 Wonderland alarm via multiplexer

Note that the inputs other than the ones for which the alarm should ring are connected to '0' and are not left unconnected because the output must be defined for every combination of inputs. The alarm should ring for six combinations and not ring for the remaining 10 combinations. As a general rule, inputs are never left unconnected, whereas outputs can be left unconnected, as they are unused (as in the decoder implementation of the circuit).

2.7 Concluding Comments and Further Reading

In this chapter we have seen the development of logic from its formal status to a circuit implementation. Many of the themes presented here will be manifest later in the book.

Two books in particular give the historical basis of logic: Boole (1958 [1854]) and Dodgson (1958 [1887]). These might be worth a look if you can find a copy. The discrete mathematics approach to logic is ably expounded by Ross and Wright (2003).

There is a plethora of texts on digital electronics. This book concentrates on principles, which do not change. The implementation technology most certainly does. Amongst those which have found popularity are: Wakerly (2005) (though at the time of writing it costs over $100 and weighs 3.6 pounds!), Floyd (2002), Katz and Borriello (2004), and Roth (2009). These are weighty texts that cover a great amount of material, but perhaps in rather excessive detail.

2.8 Questions

Solutions in Appendix 1.

Q2.1 By truth table analysis, show that

$$(A + B) \cdot \overline{A \cdot B} = A \oplus B$$
$$\overline{A + B} \cdot A \cdot B = 0$$
$$\overline{A} + \overline{B} = \overline{A \cdot B}$$
$$\overline{\overline{A + B} + \overline{A \cdot B}} = A \cdot B$$

Q2.2 Draw up a truth table for each element of the expression

$$f = \overline{A \cdot C} \cdot B \oplus D + A \cdot \overline{C} \cdot \overline{B \oplus \overline{D}} + B \cdot \overline{C} \cdot D + \overline{B + C + D}$$

and hence derive a truth table for the function f.

Q2.3 Draw the circuit to implement the function

$$f = \overline{A} \cdot \overline{B} \cdot \overline{C} + \overline{A} \cdot B \cdot \overline{C} + A \cdot B \cdot \overline{C}$$

using AND, OR, and NOT gates.

Q2.4 Implement the function

$$f = \overline{A} \cdot \overline{B} \cdot \overline{C} + \overline{A} \cdot B \cdot \overline{C} + A \cdot B \cdot \overline{C}$$

using an 3-8 line decoder and an OR gate.

Q2.5 Implement the function

$$f = \overline{A} \cdot \overline{B} \cdot \overline{C} + \overline{A} \cdot B \cdot \overline{C} + A \cdot B \cdot \overline{C}$$

using an 8-1 line multiplexer.

Q2.6 Draw the logic circuits (using AND, OR, and NOT gates) for your solutions to Q2.4 and Q2.5, and show which parts of the circuits are redundant compared with your solution to Q2.3.

Q2.7 Express the function

$$f = \overline{A} \cdot \overline{B} \cdot C + \overline{A} \cdot B \cdot \overline{C} + A \cdot \overline{B} \cdot \overline{C} + A \cdot B \cdot C$$

in software with logic functions AND, OR, and NOT applied to logic variables that can be assigned HIGH ($f = $ '1') or LOW ($f = $ '0'). Use IF...THEN...ELSE statements only.

Q2.8 Express the function in Q2.7 using the same logical variable specification, but using CASE statements only.

Q2.9 Design a safety system for a one-seater car given:

 (i) An indicator that provides an output of +5 V when someone is sat in the car and 0 V otherwise.

 (ii) An indicator that provides an output which is +5 V when the seat belt is connected properly across the occupant and 0 V otherwise.

 (iii) An indicator showing that the door is shut of +5 V output when the door is shut and 0 V otherwise.

 (iv) A system safety condition indicator for the brakes and suspension, with output +5 V when the condition is unsafe and 0 V otherwise.

The safety system should provide a signal f, which can be used to prevent the car proceeding (f = '1' to prevent progress and '0' to allow progress). The car should not proceed when a driver is not sat in it or when the driver is not wearing a seat belt or the system condition is unsafe. The car should not proceed when the door is open.

(a) Draw up a truth table for the signal f.

(b) Extract the logical function that gives the signal f.

(c) Implement the system using NOT, AND, and OR gates.

Chapter 3

Logic Circuit Implementation

3.1 Overview

So far, we have considered how to design circuits using AND, OR, and NOT gates. These certainly work correctly, but so far are rather large and need to be reduced, or minimised. To achieve this we need an algebra by which we can use logic laws to minimise the circuits. As we shall find, this becomes rather complex and so we shall then consider other minimisation tools. Throughout, we shall consider circuits and again we shall end with a practical logic design.

3.2 Laws of Logic

We have so far defined logic without a formal algebra. Including the algebra is a bit pedantic but much logic use does depend on it. Given binary variables and logical inversion, $\overline{'0'} = '1'$, $\overline{\overline{A}} = A$ we have two sets of laws, one based on OR and the other on AND.

	OR laws	AND laws
1	$A + 0 = A$	$A \cdot 0 = 0$
2	$A + 1 = 1$	$A \cdot 1 = A$
3	$A + A = A$	$A \cdot A = A$
4	$A + B = B + A$	$A \cdot B = B \cdot A$
5	$A + (B + C) = (A + B) + C$	$A \cdot (B \cdot C) = (A \cdot B) \cdot C$
6	$A + B \cdot C = (A + B) \cdot (A + C)$	$A \cdot (B + C) = A \cdot B + A \cdot C$
7	$A + B = \overline{\overline{A} \cdot \overline{B}}$	$A \cdot B = \overline{\overline{A} + \overline{B}}$

For both AND and OR, laws 1, 2, and 3 follow from the definitions of the logic functions (law 1 for AND is rather neatly called the annihilation law). Law 4 shows that it doesn't matter which way round the inputs are connected to a gate. Law 5 suggests that we can make a three-input gate from two two-input gates, and that for this function it doesn't matter how the inputs are connected, as shown in Figure 3.1, as for OR gates in which all three circuits are equivalent perform precisely the same function.

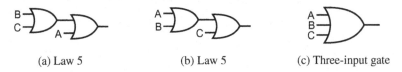

(a) Law 5 (b) Law 5 (c) Three-input gate

Figure 3.1 Equivalent circuits by the associative law, law 5

Law 6 is the distributive law, can be proved using the previous laws (it's a bit like algebraic multiplication), and is very useful for Boolean manipulation. The last two laws are versions of *De Morgan's law*. De Morgan's law shows us how to swap logic implementation.

$$\overline{A + B} = \overline{A} \cdot \overline{B} \qquad \overline{A \cdot B} = \overline{A} + \overline{B}$$

An easy way to remember how to use De Morgan's law is

Break the line and change the sign. If there is no bar, add two and break one.

The first part of the phrase is easily seen. The second part of the phrase allows use of De Morgan's law on functions which are not inverted, such as $A + B$, so:

$$A + B = \overline{\overline{A + B}} = \overline{\overline{A} \cdot \overline{B}}$$

Note that it's the lower bar that is broken (breaking the upper bar would lead to a logic function which makes no sense since inputs can only be inverted one at a time). De Morgan's law implies *universality* of NAND and NOR gates. Since both NAND and NOR gates can be used to implement OR and AND, respectively, by inverting the inputs, then any function can be implemented using only NAND or NOR gates. This is illustrated further in Section 3.5.

> *De Morgan was the first professor of maths at University College London and invented proof by induction. Not the life and soul of a party though, he was described by a colleague as a 'dry dogmatic pedant'. Note that his law is more general and you can use it to simplify complicated IF statements when programming (they are logic statements anyway!).*

De Morgan's law can also be used to explain two symbols in common use for NOR and NAND. These are due to the fact that NAND is equivalent to the OR function with inverted inputs and NOR is equivalent to AND with inverted inputs (Figure 3.2). By De Morgan's law (law 7), and by using a circle on the input connection to denote logical inversion (as in the mixed logic convention) we obtain:

$$\overline{A \cdot B} = \overline{A} + \overline{B} \quad \text{(NAND becomes OR)} \quad \text{and}$$
$$\overline{A + B} = \overline{A} \cdot \overline{B} \quad \text{(NOR becomes AND)}$$

NAND	Equivalent NAND symbol	NOR	Equivalent NOR symbol

Figure 3.2 Other gate symbols

Two other laws can be derived from this algebra:

$$A + A \cdot B = A \quad A + \overline{A} \cdot B = A + B$$

For proof, for the first of these

$$A + A \cdot B = A \cdot (1 + B) = A \cdot 1 = A$$

for the second, by substitution for $A + A \cdot B = A$ then

$$A + \overline{A} \cdot B = A + A \cdot B + \overline{A} \cdot B = A + B \cdot (A + \overline{A})$$
$$= A + B \cdot 1 = A + B$$

Formal proof is more involved and rigorous. This is a functional proof only.

These laws lead to smaller, reduced, circuits (Figure 3.3).

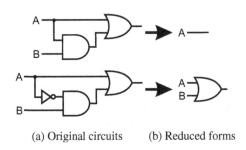

(a) Original circuits (b) Reduced forms

Figure 3.3 Circuit minimisation

So in the first case we can replace two gates with a piece of wire and in the second case we can remove two gates (and their connections) and still achieve the same function. In other words, why use those OR and AND gates when they are actually redundant? Each gate costs money and has a certain reliability and lifetime associated with it. If we can reduce the number of gates we save money and effort, and make the circuit more reliable. This is *minimisation*.

3.3 Minimisation

Circuits cost money and consume power (which also costs money!) If we reduce the circuit, we reduce the cost (increase profit) and make it more reliable. Minimisation is therefore prompted by the desire to implement a logic function in as few gates as possible to reduce cost. Designers now use logic minimisers (implemented via computer programs) that do this automatically and this will be considered later. We shall first study some by-hand techniques, to show how minimisation can be achieved. The by-hand techniques are mainly historical now, but necessary since (automated) logic minimisers develop from them. There are *algebraic* and *graphical* techniques.

3.3.1 *Algebraic minimisation*

Earlier rules give the basis for *algebraic minimisation*:

$$(i)\ A + \overline{A} = 1 \quad \text{so} \quad (ii)\ B \cdot (A + \overline{A}) = B \cdot 1 = B$$

To put the maths into words, in case *(i)* a signal OR'd with its inverted form covers all possibilities and is therefore TRUE. In case *(ii)*, a signal ANDed with an input, which in its inverted and its non-inverted form is just the signal itself (since it covers all combinations of input A). We also just demonstrated minimisation with two slightly more complicated laws:

$$A + A \cdot B = A \quad \text{and} \quad A + \overline{A} \cdot B = A + B$$

We shall use the law $B \cdot (A + \overline{A}) = B$ to minimise John Wonderland's alarm circuit. The original circuit is in Figure 2.27 and needed six four-input AND gates and one six-input OR gate. The final extraction from the truth table was:

$$f = \overline{A} \cdot B \cdot \overline{C} \cdot D + \overline{A} \cdot B \cdot C \cdot \overline{D} + \overline{A} \cdot B \cdot C \cdot D$$
$$+ A \cdot \overline{B} \cdot \overline{C} \cdot \overline{D} + A \cdot \overline{B} \cdot \overline{C} \cdot D + A \cdot \overline{B} \cdot C \cdot \overline{D}$$

which using common terms gives:

$$f = \overline{A} \cdot B \cdot D \cdot (C + \overline{C}) + \overline{A} \cdot B \cdot C \cdot (\overline{D} + D)$$
$$+ A \cdot \overline{B} \cdot \overline{C} \cdot (\overline{D} + D) + A \cdot \overline{B} \cdot \overline{D} \cdot (\overline{C} + C)$$

Note that you can consider terms more than once for the purposes of minimisation since $A + A = A$ (i.e. using a term twice does not change the function at all). In this case, we have used the two terms $\overline{A} \cdot B \cdot C \cdot D$ and $A \cdot \overline{B} \cdot \overline{C} \cdot \overline{D}$ twice, so that the minimisation can be seen. Since the terms in brackets become logic '1' the minimised result is then

$$f = \overline{A} \cdot B \cdot D + \overline{A} \cdot B \cdot C + A \cdot \overline{B} \cdot \overline{C} + A \cdot \overline{B} \cdot \overline{D}$$

which can be implemented using two-input gates by sharing common terms as

$$f = \overline{A} \cdot B \cdot (D + C) + A \cdot \overline{B} \cdot (\overline{C} + \overline{D})$$

The minimised circuits are as in Figure 3.4.

Figure 3.4 Reduced Wonderland alarm circuits

By De Morgan's law we could remove two of the inverters, since the last term can be replaced by a NAND gate as

$$f = \overline{A} \cdot B \cdot (D + C) + A \cdot \overline{B} \cdot \overline{C \cdot D}$$

We have then taken a function that originally required four inverters, six four-input gates and one six-input OR gate to seven two-input gates plus the inverters — and it still delivers the same function. By De Morgan's law, the whole circuit can be implemented using just five two-input gates and two inverters. Algebraic techniques are limited to small-scale problems; for circuits with more than five inputs they become tedious and error prone (and you might not spot the best route to minimisation). As such we need a way that is more guaranteed to provide minimised circuits.

3.3.2 *Graphical minimisation: the Karnaugh map*

The *Karnaugh map* (K-map) uses the expressions $A + \overline{A} = 1$ so $B \cdot (A + \overline{A}) = B$ in a graphical way. Essentially it is a two-dimensional truth table. We first draw a box with inputs around it, which is like a map of the logical function (Figure 3.5). The input combinations point to the value of the function for that combination.

Then we insert the value of the function for each input combination, so a K-map for AND is as per Figure 3.6.

This shows that the K-map is an alternative formulation of the truth table; it is now expressed in a compact two-dimensional way. But it has been introduced as a graphical minimisation technique, so how do we minimise using this? Consider the K-map for the OR function (Figure 3.7).

Figure 3.5 K-map basis

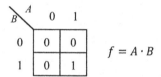

Figure 3.6 K-map for AND

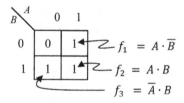

Figure 3.7 K-map for OR

By using algebraic minimisation based on f_1, f_2, and f_3 (including f_2 twice) we obtain the function f that the K-map represents as

$$f = f_1 + f_2 + f_3 = A \cdot \overline{B} + A \cdot B + \overline{A} \cdot B$$
$$= A \cdot (\overline{B} + B) + (A + \overline{A}) \cdot B$$
$$= A \cdot 1 + 1 \cdot B$$
$$= A + B$$

OK, we knew that it was OR already and that is in minimal form. The aim here is to illustrate the K-map minimisation process.

Digital Electronics: A Primer

Extraction from the truth table was based on using individual terms and then reducing them algebraically. The K-map enables this reduction to be spotted **visually** — we can see terms of the form $A \cdot B + \overline{A} \cdot B = (A + \overline{A}) \cdot B = 1 \cdot B = B$, as we just did for the OR gate. We can draw loops around these common terms to minimise the circuit. Essentially, the loop is drawn around the input that changes and then removes it (Figure 3.8).

Figure 3.8 Looping terms in a K-map

From the K-map for OR the function is then described by two loops, f_X and f_Y and since $f_X = A \cdot (\overline{B} + B) = A$ and $f_Y = (A + \overline{A}) \cdot B = B$ then the function is the sum of the two loops. The function is then $f = f_X + f_Y$, so $f = A + B$, as expected. A group of two '1's removes a single variable by covering both possible states of that variable, as in

$$f = (A + \overline{A}) \cdot B = B$$

The K-map was invented by Maurice Karnaugh in the 1950s. He was a telecommunications engineer at Bell Labs in the USA and he wanted to minimise communications circuits. Note that his use of the brain's ability to spot patterns largely preceded the developments of a modern computer.

We don't do analytic minimisation (that's far too painful), we just draw loops and that removes the unnecessary terms/inputs. To achieve the minimised circuit you should take the largest possible groups of binary size (of size 1, 2, 4, 8, and so on) to reduce the number of inputs in each

extracted term. Note that the extracted groups can overlap. The trick is to draw the loops the right way. Let's move on to some bigger K-maps. That way we can see more about the minimisation process (a two-input K-map is of zilch use, except to explain what they are). The truth table for three inputs has eight output combinations, and that for four-input combinations has 16. So their K-maps have these numbers of slots, but are more compact than truth tables since they are arranged as maps (Figure 3.9).

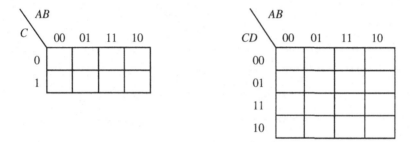

Figure 3.9 Layout of three- and four-input K-maps

Note the order of the inputs is **not** the same as for a truth table and, for two inputs, the last two columns (or rows) flip from the values expressed in the truth table. This then forces the order of the truth table to reflect the *unit distance criterion* where one input only changes between adjacent cells. If we were to order inputs as in a truth table then this grouping would not perform minimisation since in some places two inputs would change between adjacent cells so they could not be grouped for minimisation purposes.

Let's take a larger K-map (Figure 3.10) and illustrate the minimisation process by considering some of the groups within it. The unit distance criterion allows minimisation by looping the two cells in the top left corner to perform the operation

$$f_X = \overline{A} \cdot \overline{B} \cdot \overline{C} \cdot (\overline{D} + D) = \overline{A} \cdot \overline{B} \cdot \overline{C}$$

This is the loop at the top left-hand corner, loop X.

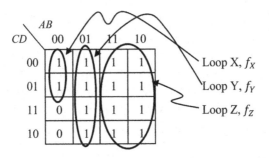

Figure 3.10 Different-sized loops

A group of four cells — loop Y — removes two variables by covering the four combinations of the inputs C and D, as

$$f_Y = \overline{A} \cdot B \cdot (\overline{C} \cdot \overline{D} + \overline{C} \cdot D + C \cdot \overline{D} + C \cdot D)$$
$$= \overline{A} \cdot B \cdot (\overline{C} \cdot (\overline{D} + D) + C \cdot (\overline{D} + D))$$
$$= \overline{A} \cdot B \cdot (\overline{C} \cdot 1 + C \cdot 1)$$
$$= \overline{A} \cdot B \cdot (\overline{C} + C)$$
$$= \overline{A} \cdot B$$

If we were to loop eight cells, we can remove three inputs, as in loop Z:

$$f_Z = A \cdot (B \cdot (\overline{C} \cdot \overline{D} + \overline{C} \cdot D + C \cdot \overline{D} + C \cdot D)$$
$$+ \overline{B} \cdot (\overline{C} \cdot \overline{D} + \overline{C} \cdot D + C \cdot \overline{D} + C \cdot D)) = A$$

Note that we just look for the terms that are not changing, rather than doing the minimisation. (That's what the K-map is for in the first place!) So by **inspection**,

$$f_X = \overline{A} \cdot \overline{B} \cdot \overline{C}, \quad f_Y = \overline{A} \cdot B \quad \text{and} \quad f_Z = A.$$

Given that the inputs are those that go through every possible combination, then we can remove loops of size 1, 2, 4, 8, and so on, in **adjacent** cells.

This begs the question: what is adjacent? The K-map wraps round side to side and top to bottom. The left column is for inputs $AB = 00$ and the rightmost column for $AB = 10$. These then still obey the unit distance criterion and were two elements in the cells they address to contain a '1',

then they can be looped around the map to reduce and remove the input A, leaving \overline{B}. The map also wraps round from top to bottom since only the input C changes between the top and bottom rows. Note that the map is not a sphere — it wraps round from top to bottom, and from side to side. Some find a K-map easier when it is drawn as in Figure 3.11, which is functionally identical to the map drawn before (Figure 3.9).

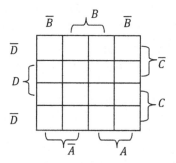

Figure 3.11 Alternative layout of a four-input K-map

3.3.3 *Examples of Karnaugh map design*

The use of the K-map for minimisation is illustrated by examples showing non-minimised and minimised solutions.

The difference in the extractions is shown by the difference between the dotted lines. Here, a larger group should have been taken in the first extraction (the left-hand side), which is reflected in the possibility of algebraically minimising the K-map result. This should not occur since the K-map is designed to avoid it. The second extraction (right-hand side) results in the minimised solution.

2)

$$f = \overline{A} \cdot \overline{C} + B \cdot C + A \cdot C$$

$$f = \overline{A} \cdot \overline{C} + \overline{A} \cdot B + A \cdot C$$

In this second example, both extractions implement the circuit in minimised form. There is no unique solution: both solutions are equally valid in terms of AND, OR, and NOT extraction.

3)

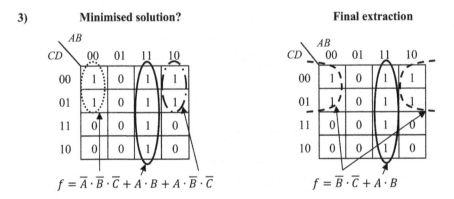

$$f = \overline{A} \cdot \overline{B} \cdot \overline{C} + A \cdot B + A \cdot \overline{B} \cdot \overline{C}$$

$$f = \overline{B} \cdot \overline{C} + A \cdot B$$

In this third extraction, we have not remembered that the table maps round from side to side and a better, more minimised extraction is possible.

4)

$$f = \overline{B} \cdot \overline{C} \cdot \overline{D} + B \cdot \overline{C} \cdot D + \overline{B} \cdot C \cdot \overline{D}$$

$$f = \overline{B} \cdot \overline{D} + B \cdot \overline{C} \cdot D$$

In this example, we have remembered that the map in its non-minimised form wraps round from side to side, but not from top to bottom. The four corner elements therefore form a group that can be extracted together as the term $\overline{B} \cdot \overline{D}$ (removing the inputs A and C).

5)

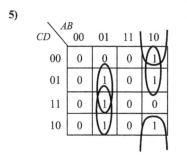

$$f = \overline{A} \cdot B \cdot C + \overline{A} \cdot B \cdot D + A \cdot \overline{B} \cdot \overline{C} + A \cdot \overline{B} \cdot \overline{D}$$

This final example is John Wonderland's alarm circuit again. Now the minimisation is clear and easy. A minimised software description for John's alarm is:

```
IF ((NOT(alice) AND bo AND (chas OR dave))
   OR (alice AND NOT(bo) AND (NOT(chas) OR NOT(dave)))
      THEN alarm=TRUE
      ELSE alarm=FALSE
```

This clearly follows the original specification. Alternatively

```
alarm = NOT(alice) AND bo AND (chas OR dave)
         OR (alice AND NOT(bo) AND (NOT(chas) OR NOT(dave))
```

This is clearly the same as the minimised solution.

3.4 Timing Considerations and Static Hazards

Any practical electronic circuit takes time to respond. For gates, the *propagation delay* defines how long before the output changes state after an input. The time for the output to change LOW to HIGH, T_{PLH}, or the time for HIGH to LOW transition, T_{PHL}, are usually of the order of nanoseconds. It is therefore important to consider how many logic gates signals pass through (how much delay there is in the circuit). This is called the *level of the logic system* (not to be confused with *logic levels* which

are the ranges of acceptable voltages for a '1' and a '0'). A two-level logic system is one where a signal will pass through two gates at maximum from the input to the output. This can be important when considering the **delay** introduced by combinational logic via T_{PHL} or T_{PLH}. A three-level logic system is shown in Figure 3.12. Here, the speed of the system is dictated by the longest path, which is the one for input B. That has, at maximum, three gates that an input must pass through, so the level of the logic system is three.

Figure 3.12 Three-level logic system

A *hazard* is clearly undesirable in a logic system. It concerns a logic variable at an **incorrect** state often due to **timing considerations** (propagation delay). The propagation delay of an inverter can cause a momentary *glitch* on a circuit output, where it has the wrong value. Consider

$$f = A \cdot \overline{B} + B \cdot C$$

We shall look at the output state for inputs $A = $ '1' and $C = $ '1' and the input B changes from '1' to '0'. We shall consider the delay from the inverter since that is what affects this circuit's operation. There is also some delay due to the AND gates and the OR gate, but that does not spell potential disaster here, whereas the NOT gate can. Essentially, the circuit output is the OR of two terms, either of which is '1' when B changes and so for $A = $ '1' and $C = $ '1' the output (in theory) should not change at all when B changes from '1' to '0'. In Figure 3.13 the horizontal axis is time and the first column is for the ideal case when \overline{B} changes to '1' precisely when the input B changes to zero. The second column is for the real case where \overline{B} changes to '1' slightly later than when the input B changes to zero.

The output staying constantly at '1' is true in theory, but in practice \overline{B} is the output of an inverter whose input is B. When B changes from '1' to '0' then the output of the inverter, \overline{B}, changes from '0' to '1' T_{PLH} ns

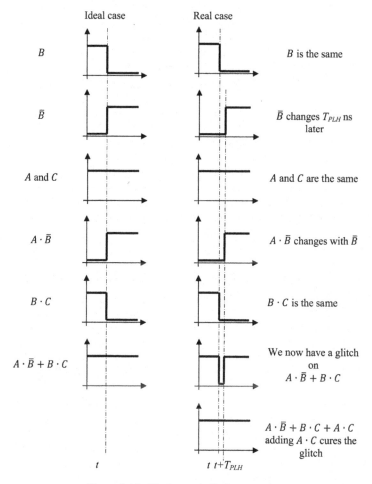

Figure 3.13 Timing and glitch generation

later. There is then a short time during which both $A \cdot \overline{B}$ and $B \cdot C$ are both zero. The output will then be momentarily '0' whilst the change in B propagates through the inverter. This is unwelcome but it is realistic, since no circuit can respond infinitely fast; this is called a *static hazard*. To reduce its effect we can link together all the largest groups extracted for minimisation. The extra terms we introduce are called *bridging terms*. The K-map for $f = A \cdot \overline{B} + B \cdot C$ (Figure 3.14) shows how the bridging term is derived.

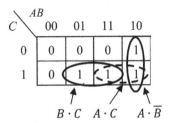

Figure 3.14 Using a bridging term

The bridging term $A \cdot C$ is not affected by the change in \overline{B}, and its inclusion in the overall function removes the glitch due to the static hazard. The hazard-free implementation is then

$$f_{HF} = B \cdot C + A \cdot C + A \cdot \overline{B}$$

Some further examples show K-maps where bridging terms are not possible, and where they are possible (Figure 3.15).

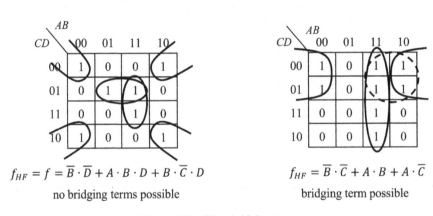

$f_{HF} = f = \overline{B} \cdot \overline{D} + A \cdot B \cdot D + B \cdot \overline{C} \cdot D$

no bridging terms possible

$f_{HF} = \overline{B} \cdot \overline{C} + A \cdot B + A \cdot \overline{C}$

bridging term possible

Figure 3.15 Other bridging terms

In practice, the treatment of hazards is more complex. There are some hazards that are not exposed by determining the bridging terms in K-maps and it is often not possible to link all terms together. It is often better to **wait** until any timing components should have settled (i.e. expect propagation delay as part of life and design around it). The time you should wait for is given by the worst-case path through the logic (i.e. the path that a signal takes the longest time to traverse).

3.5 Logic Implementations

3.5.1 *Using NAND or NOR to provide NOT*

In Section 3.2 it was stated that De Morgan's law implies that NAND and NOR gates are universal by virtue of the possibility of exchanging AND/OR function. In fact, in a more general sense universality exists so long as a NOT gate is available. Given that for universality we want to build a circuit using NAND (or NOR) gates only, we then need to determine whether we can derive the NOT function using a NAND gate (or a NOR gate). This can be achieved by study of the truth tables (Table 3.1).

Table 3.1 Inversion by NAND and NOR

A	B	$\overline{A \cdot B}$	$\overline{A + B}$
0	0	1	1
0	1	1	0
1	0	1	0
1	1	0	0

By inspection we can see that the output of the NAND gate $\overline{A \cdot B} = \overline{A}$ if the input B equals A (i.e. the two inputs are connected together), or if $B = $ '1' then the output is the inverted form of the input A. Similarly, the output of the NOR gate $\overline{A + B} = \overline{A}$ if $B = $ '0' or when $B = A$, giving four possible circuits equivalent to NOT (Figure 3.16).

Figure 3.16 Other inverter circuits

The circuits where one input is connected low or high are usually preferred since the input is connected to a power supply, not to the output of another gate. This takes current from the power supply (which is why it is there) rather than from the output of another gate. (Note that the consequences of interconnection on logic level are handled in greater detail in Chapter 4 where we consider the circuits that implement the gates.)

3.5.2 *Extracting '0's or '1's?*

The notion of universality can be further investigated by considering a K-map, wherein we have so far considered extraction by the '1's only.

Naturally there are '0's and these can be grouped to determine the **inverted** function. In a K-map, extracting the '1's gives an AND/OR function whereas by extracting the '0's we achieve AND/NOR. To show this, and in part to try to answer the question as to which is best, we shall start with an example K-map (Figure 3.17).

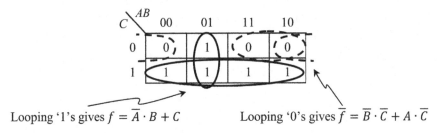

Looping '1's gives $f = \overline{A} \cdot B + C$ Looping '0's gives $\overline{f} = \overline{B} \cdot \overline{C} + A \cdot \overline{C}$

Figure 3.17 Example K-map

By **extracting the '1's** we achieve

$$f = \overline{A} \cdot B + C$$

This can be implemented using NAND gates only by application of De Morgan's law, thereby illustrating the universality of NAND gates: the circuit can be implemented without AND and without OR, just NAND.

$$f = \overline{\overline{\overline{A} \cdot B} \cdot \overline{C}}$$

This requires two two-input NAND gates (plus two NAND gates to invert A and C as in Section 3.5.1).

By **extracting the '0's** (by looping the '0's in the same way we loop the '1's) we extract the inverse function \overline{f}

$$\overline{f} = \overline{B} \cdot \overline{C} + A \cdot \overline{C}$$

and by inverting both sides

$$f = \overline{\overline{B} \cdot \overline{C} + A \cdot \overline{C}}$$

This is an AND/NOR function. We can again achieve an entirely NOR construction by applying De Morgan's law, re-illustrating that NAND and

NOR gates are both universal:

$$f = \overline{\overline{B + C} + \overline{\overline{A} + C}}$$

This requires three two-input NOR gates, plus one NOR gate to invert A. To demonstrate the duality between extraction by the '0's and by the '1's, by applying De Morgan's law again

$$f = \overline{\overline{B + C}} \cdot \overline{\overline{\overline{A} + C}}$$

which gives

$$f = (B + C) \cdot (\overline{A} + C)$$

by expansion we obtain

$$f = \overline{A} \cdot B + \overline{A} \cdot C + B \cdot C + C \cdot C$$

and by minimisation

$$f = \overline{A} \cdot B + C$$

This returns us to the function extracted by the '1's, hence showing the duality of the two extractions. Which is best? That rather depends on an overall system. In this case extracting the '1' leads to one gate less, but that is not always the case.

'Contrariwise,' continued Tweedledee, 'if it was so, it might be, and if it were, it would be: but as it ain't, it ain't. That's logic.' (Through the Looking Glass, Lewis Carroll, Penguin, 2007)

3.6 Terminology

The extraction of a function by grouping inputs using AND gates whose outputs are grouped using OR is called the *sum of products* (SOP) form, e.g.

$$f = \overline{A} \cdot \overline{B} \cdot \overline{C} + A \cdot B \cdot \overline{C}$$

It conforms rather nicely with the way we think about things. There is conversely a *product of sums* (POS), e.g.

$$f = (\overline{A} + B) \cdot (A + C)$$

A product term containing one occurrence of **every** variable is known as a *minterm* or a *canonical product term*. A sum term containing one occurrence of **every** variable is known as a *maxterm* or a *canonical sum term*. A function is expressed as OR of distinct minterms is known as *canonical sum of products*; one expressed as AND of distinct maxterms is known as *canonical product of sums*. For those with masochistic tendencies, canonical SOP is *disjunctive normal form* and canonical POS is *conjunctive normal form*. For further developments let us consider a function given by a truth table in Table 3.2.

Table 3.2 Example logic function showing binary truth table index

Minterm number	J	K	L	f
0	0	0	0	0
1	0	0	1	0
2	0	1	0	0
3	0	1	1	1
4	1	0	0	1
5	1	0	1	1
6	1	1	0	1
7	1	1	1	1

Here the decimal number corresponds to the binary representation of the input coding — its minterm number. This function can be expressed using a decimal index for each term in the truth table. The minterms are the cases for which the function is '1' and collecting these in SOP form is

$$f = \overline{J} \cdot K \cdot L + J \cdot \overline{K} \cdot \overline{L} + J \cdot \overline{K} \cdot L + J \cdot K \cdot \overline{L} + J \cdot K \cdot L$$

which can be expressed in a shortened manner using the decimal index as the canonical SOP

$$f = \sum_{JKL} (3, 4, 5, 6, 7)$$

These indices are a function of expressing *JKL* as a binary number where *J* is the most significant bit (MSB) and *L* is the least significant bit (LSB),

as in Table 3.2. The function can also be expressed as a product of sums

$$f = (J + K + L) \cdot (J + K + \overline{L}) \cdot (J + \overline{K} + L)$$

And in shortened form, the canonical POS is

$$f = \prod_{JKL}(0, 1, 2)$$

So how are these related? By extracting the '0's

$$\overline{f} = \overline{J} \cdot \overline{K} \cdot \overline{L} + \overline{J} \cdot \overline{K} \cdot L + \overline{J} \cdot K \cdot \overline{L}$$

By inversion

$$f = \overline{\overline{J} \cdot \overline{K} \cdot \overline{L} + \overline{J} \cdot \overline{K} \cdot L + \overline{J} \cdot K \cdot \overline{L}}$$

Now, by De Morgan's law

$$f = \overline{\overline{J} \cdot \overline{K} \cdot \overline{L}} \cdot \overline{\overline{J} \cdot \overline{K} \cdot L} \cdot \overline{\overline{J} \cdot K \cdot \overline{L}}$$

Again, by De Morgan's law

$$f = (J + K + L) \cdot (J + K + \overline{L}) \cdot (J + \overline{K} + L)$$

Phew, we're back to the expression of product of sums. This is the canonical product of sums specification. Note the 'inversion' in canonical POS expression and the relation between SOP and POS. You often get logic functions expressed in POS and SOP form (using summation or product symbols, respectively) — it is a very compact description, and one not prone to misinterpretation.

Some of these terms might seem a bit pedantic and tortuous. However, they do allow for unambiguous and compact specification, and so are seen in professional use. Some terms found in some textbooks have been omitted here, as they are rather unnecessary, e.g. an 'isolated — cell' is what it says, it's a canonical SOP term for which no minimisation can be achieved. Erm, obvious?!

By K-map for the original function f, see Figure 3.18.

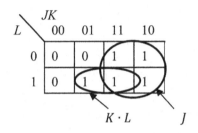

Figure 3.18 Example K-map for the original function f

So $f = J + K \cdot L$ is the condensed or minimised form (i.e. a *minimal sum*). The terms J and $K \cdot L$ are non-canonical (they do not contain all the variables) and are called *prime implicants*; these are the largest loops you can draw in the K-map. For a K-map containing four prime implicants, the four could be included in an implementation aimed to avoid static hazards (Figure 3.19).

Figure 3.19 Prime implicants and essential prime implicants

Note that the minimal sum does not necessarily contain all prime implicants, but a hazard-free one (probably) does. Terms that must be included in the minimal sum are called *essential prime implicants* (EPIs; these are the dotted groups in Figure 3.19) and are the only prime implicants that cover a particular '1' cell. Other '1's may be covered by two or more prime implicants, **one** of which must be included in the minimal sum.

3.7 Design Example

Let us consider the following design example. Essentially, it concerns designing a circuit to compare the magnitude of two binary numbers and to provide signals indicating the results.

A circuit is required which will compare the magnitude of two two-bit binary numbers, *AB* and *CD*. Each number may have a value 0 (when *A* = '0', *B* = '0' and *C* = '0' and *D* = '0'), 1 (when *A* = '0', *B* = '1' and *C* = '0', *D* = '1'), 2 (when *A* = '1', *B* = '0' and *C* = '1', *D* = '0'), or 3 (when *A* = '1', *B* = '1' and *C* = '1', *D* = '1'). The complemented outputs of *A*, *B*, *C*, and *D* are available. Circuit outputs *G*, *E*, and *L* are to be TRUE ('1') when the number *AB* is greater than, equal to, or less than the number *CD*, respectively. The outputs should be FALSE ('0') otherwise.

(a) Produce a K-map for each output (*G*, *L* and *E*).
(b) Implement *G*, *L*, and *E* using a minimum number of AND, OR, or NOT gates.
(c) Implement *G* using NAND gates only.
(d) Implement *E* using NOR gates only.
(e) Show how *L* may be obtained using an 8–1 multiplexer.
(f) Determine the relationship between *G* and the function

$$f = A \cdot \overline{C} \cdot (B \oplus D) + \overline{A + B + C + D} + A \cdot \overline{C} \cdot D + B \cdot D \cdot \overline{A \oplus C}$$

(a) The K-maps are:

CD\AB (G)	00	01	11	10
00	0	1	1	1
01	0	0	1	1
11	0	0	0	0
10	0	0	1	0

CD\AB (E)	00	01	11	10
00	1	0	0	0
01	0	1	0	0
11	0	0	1	0
10	0	0	0	1

CD\AB (L)	00	01	11	10
00	0	0	0	0
01	1	0	0	0
11	1	1	0	1
10	1	1	0	0

As a cross-check, if we were to place these K-maps on top of each other, we would find that every box has just one '1'. That's exactly as the

specification. If the specification had been $G = $ '1' for $AB \geq CD$ or $L = $ '1' for $AB \leq CD$ then this would not have been the case. The K-map for E is routine: each occupied cell is where the inputs are the same and this must be the diagonal.

(b) The minimal extraction for each is by the prime implicants.

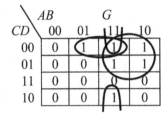

$$G = A \cdot \overline{C} + B \cdot \overline{C} \cdot \overline{D} + A \cdot B \cdot \overline{D}$$

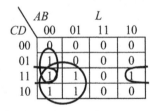

$$E = \overline{A} \cdot \overline{B} \cdot \overline{C} \cdot \overline{D} + \overline{A} \cdot B \cdot \overline{C} \cdot D +$$
$$A \cdot \overline{B} \cdot C \cdot \overline{D} + A \cdot B \cdot C \cdot D$$

CD \ AB	00	01	11	10
00	0	0	0	0
01	1	0	0	0
11	1	1	0	1
10	1	1	0	0

L

$$L = \overline{A} \cdot C + \overline{B} \cdot C \cdot D + \overline{A} \cdot \overline{B} \cdot D$$

Note that there is some symmetry between the expressions for G and for L, which is to be expected. Extracting E from a K-map appears to be overkill and we could write the expression by inspection. As usual, we often find this out afterwards.

We can check these circuits using Logisim:

Here, the function L has been implemented both by the K-map extraction and by combining G and E. This is by NOR since L is not G, AND not E, and by De Morgan's law this becomes a NOR function.

$$L = \overline{G} \cdot \overline{E} = \overline{G + E}$$

(c) To implement G using NAND gates only we use De Morgan's law:

$$G = A \cdot \overline{C} + B \cdot \overline{C} \cdot \overline{D} + A \cdot B \cdot \overline{D} = \overline{\overline{A \cdot \overline{C}} \cdot \overline{B \cdot \overline{C} \cdot \overline{D}} \cdot \overline{A \cdot B \cdot \overline{D}}}$$

This can be achieved using NAND gates only since we can use NAND to give NOT by holding one input high $\overline{A \cdot 1} = \overline{A}$. If we needed a minimised implementation we would need to consider the K-map, but that is not the case here.

(d) To implement E using NOR gates we can either twice invert the function already extracted (since a NOR gate with one input LOW acts as an inverter)

or extract the '0's. By inverting twice the previous canonical SOP extraction for E we obtain

$$E = \overline{\overline{\overline{A} \cdot \overline{B} \cdot \overline{C} \cdot \overline{D} + \overline{A} \cdot B \cdot \overline{C} \cdot D + A \cdot \overline{B} \cdot C \cdot \overline{D} + A \cdot B \cdot C \cdot D}}$$

Then by inverting each term twice (ready for De Morgan's law) we obtain

$$E = \overline{\overline{\overline{\overline{A} \cdot \overline{B} \cdot \overline{C} \cdot \overline{D}}} + \overline{\overline{\overline{A} \cdot B \cdot \overline{C} \cdot D}} + \overline{\overline{A \cdot \overline{B} \cdot C \cdot \overline{D}}} + \overline{\overline{A \cdot B \cdot C \cdot D}}}$$

So by De Morgan's law

$$E = \overline{\overline{\overline{A + B + C + D}} + \overline{A + \overline{B} + C + \overline{D}} + \overline{\overline{A} + B + \overline{C} + D}}$$
$$+ \overline{\overline{A} + \overline{B} + \overline{C} + \overline{D}}$$

And this uses NOR gates only. Alternatively, by extracting the '0's

$$\overline{E} = B \cdot \overline{D} + A \cdot \overline{C} + \overline{A} \cdot C + \overline{B} \cdot D$$

by applying De Morgan's to each term

$$\overline{E} = \overline{\overline{B} + D} + \overline{\overline{A} + C} + \overline{A + \overline{C}} + \overline{B + \overline{D}}$$

and by inverting each side

$$E = \overline{\overline{\overline{B} + D} + \overline{\overline{A} + C} + \overline{A + \overline{C}} + \overline{B + \overline{D}}}$$

and this gives a much smaller circuit.

(e) To implement L using an 8–1 multiplexer we can view the function in non-minimised form as

$$L = \overline{A} \cdot \overline{B} \cdot \overline{C} \cdot D + \overline{A} \cdot \overline{B} \cdot C \cdot \overline{D} + \overline{A} \cdot \overline{B} \cdot C \cdot D + \overline{A} \cdot B \cdot C \cdot D$$
$$+ \overline{A} \cdot B \cdot C \cdot \overline{D} + A \cdot \overline{B} \cdot C \cdot D$$

This can be implemented in a multiplexer with control lines BCD addressing the eight channels. For the channels addressed by the first four terms, corresponding to multiplexer input channels with $BCD = 001$, 010, 111, and 110, respectively, the multiplexer inputs can be fed with the inverted input \overline{A}, since this is common to all four terms. For the last term in L the output is '1' if $B = $ '0', $C = $ '1', and $D = $ '1', and so we connect a high input

to the multiplexer input addressed by that term. All other input channels addressed by uncovered combinations of *BCD* (in the expression for *L*) are connected to '0'.

If we were required to use a smaller multiplexer then more logic would be used at the input stage; should a larger multiplexer be used (i.e. a 16–1 line multiplexer) then the inputs would become '0' or '1' and the input combinations would select chosen logic levels, those specified in the K-map for *L*.

(**f**) Essentially, we use logic laws to determine a sum of products and we use this to complete a K-map:

$$f = A \cdot \overline{C} \cdot (B \oplus D) + \overline{\overline{A} + B + C + D} + A \cdot \overline{C} \cdot D + B \cdot D \cdot \overline{A \oplus C}$$

By the function of XOR and XNOR

$$f = A \cdot \overline{C} \cdot (B \cdot \overline{D} + \overline{B} \cdot D) + \overline{\overline{A} + B + C + D} + A \cdot \overline{C} \cdot D$$
$$+ B \cdot D \cdot (\overline{A} \cdot \overline{C} + A \cdot C)$$

And then by expanding these terms

$$f = A \cdot B \cdot \overline{C} \cdot \overline{D} + A \cdot \overline{B} \cdot \overline{C} \cdot D + \overline{\overline{A} + B + C + D} + A \cdot \overline{C} \cdot D$$
$$+ \overline{A} \cdot B \cdot \overline{C} \cdot D + A \cdot B \cdot C \cdot D$$

By De Morgan's law

$$f = A \cdot B \cdot \overline{C} \cdot \overline{D} + A \cdot \overline{B} \cdot \overline{C} \cdot D + A \cdot \overline{B} \cdot \overline{C} \cdot \overline{D}$$
$$+ A \cdot \overline{C} \cdot D + \overline{A} \cdot B \cdot \overline{C} \cdot D + A \cdot B \cdot C \cdot D$$

Then use these terms to make up a K-map for f — it is the same as the one for G.

3.8 Algorithmic Minimisation (Quine–McCluskey and Espresso)

Obviously, minimisation is tedious, repetitive, and prone to error. That's what computers are good at, and that's why they are great at minimising logic circuits. Obviously, we need to enter the circuit data in some way into a computer to achieve this, and in modern designs this is achieved by a software specification. We shall start with the *Quine–McCluskey method*, which is a method of prime implicants. The first stage is to determine all the prime implicants, then we determine the smallest set of prime implicants that satisfy the circuit's function, thereby minimising it. This set must include all the prime implicants. We shall illustrate the procedure using Wonderland's alarm from Section 2.6, which was

$$alarm = \overline{A} \cdot B \cdot \overline{C} \cdot D + \overline{A} \cdot B \cdot C \cdot \overline{D} + \overline{A} \cdot B \cdot C \cdot D$$
$$+ A \cdot \overline{B} \cdot \overline{C} \cdot \overline{D} + A \cdot \overline{B} \cdot \overline{C} \cdot D + A \cdot \overline{B} \cdot C \cdot \overline{D}$$

and this was minimised in Section 3.3.3 as

CD \ AB	00	01	11	10
00	0	0	0	1
01	0	1	0	1
11	0	1	0	0
10	0	1	0	1

$$alarm = \overline{A} \cdot B \cdot C + \overline{A} \cdot B \cdot D + A \cdot \overline{B} \cdot \overline{C} + A \cdot \overline{B} \cdot \overline{D}$$

The minterms are

$$alarm = \sum_{ABCD} (5, 6, 7, 8, 9, 10)$$

Listing these minterms in logic form we have

#	A	B	C	D
5	0	1	0	1
6	0	1	1	0
7	0	1	1	1
8	1	0	0	0
9	1	0	0	1
10	1	0	1	0

We then work out how the groups can combine, developing a list of terms that can be combined (algorithmically mimicking groups of two in the K-map) and replacing the bit that can be ignored with an X. We then get

Terms	A	B	C	D
(5,7)	0	1	X	1
(6,7)	0	1	1	X
(8,9)	1	0	0	X
(8,10)	1	0	X	0

We then repeat this iteration as far as we can (first by looking for groups of four in the K-map, and then 16 et seq.). In this case, we do not have any further groups so our minimised function is

$$alarm = 01X1 + 011X + 100X + 10X0 = \overline{A} \cdot B \cdot D + \overline{A} \cdot B \cdot C$$
$$+ A \cdot \overline{B} \cdot \overline{C} + A \cdot \overline{B} \cdot \overline{D}$$

Published papers can include a photograph of the authors (if provided). McCluskey always favours a different hat in these photos (McCluskey, 2001). Well done, Ed!

This procedure can be automated and then implemented as a computer algorithm. This procedure will clearly explode when there are more inputs.

If $n = 32$ there could be more than 6.5×10^{15} prime implicants. OK, storage is cheap now, and so is processing power. That tends to mean our designs are more complex so we need an optimised search procedure. The one that everyone uses is called the *Espresso* system, and it came from IBM in the 1980s. This is heuristic, but operates well. Essentially, it looks for the ON and for the OFF terms which best minimise the function. The algorithm finds pretty well standard usage, and there are several variants. Well done, Robert Brayton: as ever history tends not to record gratitude or honours for engineers (remember from Chapter 1 an engineer called Shannon who invented computers). Brayton has plenty of academic awards, but as ever, people probably know the name 'Espresso' better than 'Brayton'.

3.9 Concluding Comments and Further Reading

In this chapter we have seen the development of logic from its specification status to a circuit implementation. Many of the themes presented in this chapter will be manifest later in this book. The development of minimisation has studied some by-hand techniques and some of the algorithmic techniques. These techniques actually develop into computer programs that can be used for circuits with large numbers of inputs. The K-map is very restricted in this respect. Imagine drawing a K-map for a function of ten inputs with 1,024 spaces, and then you have to start finding the loops! It is sufficient as a hand tool and gives useful insight, but there its utility ends. But then again, the algebraic techniques really need a computer. Finally, we have seen some of the basic building blocks and been introduced to the design of combinational logic systems. These will be used throughout the book, particularly in the design examples. In fact, they are a notational convenience, a sort of functional shortcut in the design process. In practical designs, these will be subject to minimisation. However it is worth noting that regular layout can lead to more efficient minimisation in very-large-scale integration (VLSI) than can be achieved with a fully minimised deign. In this respect, design and minimisation are topics of continuing interest and research (see, for example, Brayton *et al.* [1984] [you just saw this name above. . .]). For the academics (historians?), try Karnaugh (1953); the *Espresso* method is found with McGeer *et al.* (1993); for Quine–McCluskey see McCluskey (1956). Another topic of major interest is the specification

of circuits and their operation. A software specification clearly expresses this, which is itself another avenue for further investigation.

3.10 Questions

Solutions in Appendix 1.

Q3.1 Implement the function

$$f = \sum_{ABCD}(4, 6, 9, 11)$$

using a minimum number of standard logic gates (AND, OR, NOT, NAND, or NOR) that can have any number of inputs. Implement it again using NOR gates only.

Q3.2 Implement the function

$$f = \prod_{ABCD}(0, 2, 4, 5, 6, 7, 11, 15)$$

assuming that complemented inputs are available, but without static hazards, and using a minimum number of standard logic gates (AND, OR, NOT, NAND, or NOR) that can have any number of inputs.

Q3.3 Implement the function

$$f = \overline{B} \cdot \overline{C} \cdot \overline{D} + \overline{B} \cdot \overline{C} \cdot \overline{A \cdot D} + A \cdot \overline{B} \cdot \overline{C} + B \cdot C \cdot A \oplus D$$
$$+ \overline{A} \cdot B \cdot C + B \cdot C \cdot \overline{\overline{A} \cdot \overline{D}}$$

using a minimum number of standard logic gates (AND, OR, NOT, NAND, or NOR) and express your solution in sum of products form.

Q3.4 Implement the function

$$f = \overline{B} \cdot \overline{D} \cdot \overline{A \oplus C} + B \cdot D \cdot \overline{A \oplus C} + A \cdot C \cdot \overline{\overline{B} \cdot \overline{D}}$$
$$+ \overline{B} \cdot (A \cdot C + \overline{A} \cdot \overline{C}) + \overline{A} \cdot \overline{C} \cdot \overline{B} \cdot D$$

using a minimum number of standard logic gates (AND, OR, NOT, NAND, or NOR) and express your solution in product of sums form.

Q3.5 What is the level of logic associated with the function

$$f = \overline{A \cdot \overline{B} \cdot D + A \cdot C \cdot \overline{D}}$$

when implemented using standard logic gates with any number of inputs? If the T_{PLH} for each combinational logic gate is 10 ns and T_{PHL} for each combinational logic gate is 11 ns, what is the worst-case propagation delay through the circuit?

Q3.6 Identify the static hazards which might be associated with a direct implementation of the following expression:

$$f = \overline{A} \cdot C \cdot D + B \cdot C \cdot \overline{D} + A \cdot \overline{C} \cdot D + A \cdot \overline{B} \cdot \overline{C} \cdot \overline{D}$$

Using a K-map show how these static hazards might be removed. The function is required as part of a control system and the designer needs to know how long after the inputs have changed the output can be used. For all gates T_{PHL} is 9 ns and T_{PLH} is 10 ns. By considering the level of logic, estimate how long after the inputs change that the output of the original function and the output of your hazard-free one can be used by the designer.

Chapter 4

Logic Circuits

4.1 Logic Circuit Characteristics

This chapter describes what is inside an integrated circuit. Integrated circuits are made from *switching devices*. These switching devices connect together inputs to implement logic functions. An ideal switching device would respond infinitely fast, consume no power, and be physically extremely small. This is practically impossible to achieve and any switching circuit is a compromise between these factors. The *logic technology* concerns the physical implementation of the switching circuits and offers a distinct set of *performance characteristics*. We shall first investigate the performance characteristics of particular interest to integrated circuit technology, then go on to look at switching circuits, and finally consider the logic technologies that are currently available.

4.1.1 *Transfer characteristics*

The performance characteristics can be developed from an **ideal** logic gate. The ideal inverter connects its output HIGH to the positive supply rail for a LOW input. For a HIGH input the output is connected LOW to the negative voltage supply, usually ground. The input then acts adversely on a pair of switches; one connects the output HIGH, the other connects the output LOW, as shown in Figure 4.1.

Figure 4.1 Ideal inverter

Since the input acts adversely on either switch, the upper switch is turned ON for a LOW input, which is signified by the inversion symbol attached to the input to the switch. Further gates, such as AND and NAND, require appropriate combination of the input signals to control the output switches and an ideal logic gate is shown in Figure 4.2.

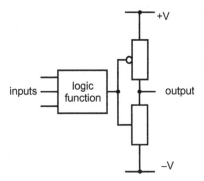

Figure 4.2 Ideal combinational logic gate

Ideally, the switches connect the output (voltage) level HIGH or LOW in response to an input voltage level. We are then considering *level logic* where voltage levels are used to signify logic variables. In this introduction we shall use *positive logic*, which is a convention that a HIGH voltage level, usually +5 V, represents a logic '1'. Accordingly, in positive logic a LOW level, usually 0 V, signifies a logic '0'. The *transfer characteristic* shows the relationship between inputs and outputs, and for (shown in Figure 4.3[a])

an **ideal** inverter the transfer relationship has only two possible values for the output. The slope of the line between the two output states is vertical, which implies that the ideal inverter has an infinite gain in the transition region.

(a) Ideal inverter transfer characteristic (b) Real inverter transfer characteristic

Figure 4.3 Transfer characteristics

This gain is impossible to achieve in practice. Also, the use of a single voltage level to signify a '1' or a '0' is very difficult to achieve. Most circuits use a range of voltage levels to represent logic signals. The transfer characteristic then relates ranges of input and output voltage levels representing '1' and '0'. The transfer characteristic of a **real** inverter then differs from that of its ideal counterpart, as shown in Figure 4.3(b).

> *Manufacturers have to guarantee a device's performance limits. These include the electrical and switching characteristics, which can be thought of as DC and AC parameters, respectively.*

For the HIGH state, the logic levels most commonly encountered in logic design range from 2.0 V to 5.0 V and for a LOW state the level is in between 0 V to 0.8 V to indicate a '0'. This depends on the power supplies that are used. Any electronic circuit requires a positive and negative power supply and for a switching circuit these supplies only need to differ in magnitude. The power supply clearly affects the voltage levels used to represent a '1' and a '0'. Should two logic technologies use different

power supplies then the logic levels in each will be different and if the two technologies need to be connected together (i.e. a circuit has been designed to take advantage of particular performance characteristics of each logic technology) then this will pose difficulty in *interfacing* (connecting together) the two circuits.

4.1.2 *Performance characteristics*

Manufacturers specify *guaranteed output levels* and *specified input levels*. This is a guarantee that if an input is within the range specified for the input levels, then the output of the circuit is guaranteed to be within a range specified for the output voltage levels. If you buy a circuit that does not conform to this specification you can return it to the manufacturer for replacement.

If an input voltage level falls outside the range specified for a valid logic level then it naturally becomes invalid and the circuit will not operate correctly. As in Figure 4.4 this might occur because noise has corrupted the transmission channel connecting two logic gates.

Figure 4.4 Noise in transmission channels

Noise can be induced by electromagnetic radiation from an external source of sufficient magnitude to force the voltage level on the transmission channel to become invalid. In reality we actually live in a fog of electromagnetic radiation and it is unsurprising to find that some gets picked up in logic circuits. The designers of integrated circuits accommodate this by introducing a *noise margin* into a circuit; this is a measure of by how much a valid signal can be corrupted by noise and still be interpreted correctly by the logic circuit. This can be summarised by relating the output levels to the input voltage levels.

The noise margin for a HIGH level, N_{MH}, is a measure of how much noise can be added to a signal that will still be interpreted as a valid '1', as shown in Figure 4.5. It is the difference between the minimum level for a valid output '1' and the (specified) minimum level for a valid input '1'. Note that in the HIGH state we are concerned with noise that **reduces** the output level. The noise margin for a LOW level, N_{ML}, concerns noise that **increases** an output level and is equal to the difference between a valid output '0' and a valid input '0'. N_{ML} is then a measure of how much noise can be added to a LOW logic signal whilst it is still interpreted correctly by later combinational circuitry. Logic technologies usually specify a noise margin, which is equal in both high and low states. There is an uncertainty region between the maximum value for a LOW input and the minimum value for a HIGH input; if a circuit input falls within the uncertainty region then the output is not guaranteed by the manufacturer. This situation should be avoided since it can result in unreliable circuit operation. It is clearly desirable to have the largest possible noise margin, but some logic technologies can offer only a small noise margin while possessing other attributes.

Figure 4.5 Noise margins

The logic level might become invalid because of contamination by noise. It might also become invalid because an output has been connected to too many inputs. The maximum number of gate inputs that can be connected to the output of a single gate is usually expressed as the *fan-out* of that gate.

> *Some texts also refer to 'fan-in' and there are a number
> of conflicting definitions. The most convincing definition
> of fan-in is the maximum number of logic inputs that a
> gate can accept. (One egregious definition is that it is the
> number of input pins — and that's plain daft!)*

Fan-out can be explained in terms of valid logic levels. In an ideal gate a HIGH output is connected to the positive supply through a perfect switch. A perfect switch is one that has no resistance when it is switched ON (a short circuit) and infinitely high resistance (an open circuit) when it is switched OFF. When it is ON there is no voltage drop across it since it has no resistance. Logic circuits do not use perfect switches; they use switches which **approximate** the perfect switch. Since they only approximate their perfect version they have resistance. Their resistance implies that the voltage across them increases with an increase in the current through them. When the voltage drop increases the output voltage falls, and when the output is connected to too many inputs the voltage will fall beneath that guaranteed as a valid output. The increase in current is due to increasing the number of gates connected to the output. Should we connect more gates than the specified maximum (the fan-out of the device), the output voltage level will not be valid as specified for the device.

If the difference between logic levels is small then the noise margin is bound to be very small as well. The advantage associated with small difference between logic levels is **speed**, since the logic technology has only to switch the output by a small amount. Speed is usually expressed in terms of *propagation delay*, which is a measure of how long a change in the inputs takes to propagate through to the circuit output. The propagation delay can differ for an output changing from LOW to HIGH or from HIGH to LOW. T_{PLH} measures the time between a change in inputs and the subsequent change in output from '0' to '1'. T_{PHL} is a measure of the time taken for the outputs to change from HIGH to LOW, '1' to '0', in response to a change in the inputs. This is the time that gave rise to the static hazard in Section 3.4. The timing is illustrated in Figure 4.6 where the time is measured from when

the inputs changed. The duration is measured to the 50% point, where the output is halfway between its initial and its final states.

The propagation delay largely arises from the need to shift charge and that takes time; less to shift means less time is needed.

Figure 4.6 Propagation delays

It is advantageous to have the fastest possible switching time. This will usually incur a higher *power consumption* since this is inherent in a switching process. **Speed** and **power consumption** are often the most important factors when considering different logic technologies. They are often combined in a *speed/power product* which is formulated by multiplying the worst-case propagation delay (the longest time taken by a device to respond) by the power consumed on average by the device. This can then be used as a performance metric to compare logic technologies.

The **main** criteria for a logic technology are then the speed it can achieve (the smallest propagation delay) and the power consumed by the device. The secondary performance characteristics are the logic levels, noise margin, and fan-out. The **secondary** criteria might dominate choice in a particular application since they dominate circuit implementation. All of these factors are a compromise; high speed often incurs high power consumption, good fan-out usually implies wide noise margins but since the logic levels are wide the speed is then reduced. Any logic technology offers a compromise between these factors and this compromise depends on the switching elements used to implement the logic function, together with implementation criteria, in particular the logic levels.

4.2 Switching Devices

4.2.1 *The diode*

The simplest form of electronic switch is a *semiconductor diode*. This is essentially a switch that can be turned ON or OFF. Its symbol indicates the direction in which current can pass through it (Figure 4.7).

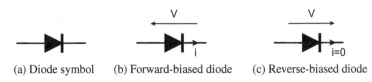

(a) Diode symbol (b) Forward-biased diode (c) Reverse-biased diode

Figure 4.7 The diode

If we apply a voltage consistent with the current direction (the positive connection, the anode, is more positive than the negative connection, the cathode) then current can pass through the device and it is switched ON: in this condition the diode is termed *forward biased*. If we apply a voltage the other way round (the cathode is more positive than the anode) no current can flow and the diode is switched OFF; in this condition the diode is termed *reverse biased*. The *equivalent circuit* is a model of the circuit and for a forward-biased diode it is a short circuit. An equivalent circuit for a reverse-biased diode is an open circuit (since no current can pass). There is actually a **threshold** voltage for a forward-biased diode; the threshold voltage must be exceeded before current can pass and this is called the *cut-in voltage* (or *knee voltage*). This can be seen from the current/forward voltage characteristic, which illustrates that a diode does not conduct before the threshold voltage is reached; after the threshold voltage is exceeded, the current increases rapidly.

> *One of the oldest canards about electronics is that it is the only subject whose name derives from something we have never seen (OK, some, like historians, will disagree). If we could see an electron, what would it look like?*

The diode is made by joining two types of semiconductor, shown in Figure 4.8. One of these is *n-type semiconductor* (negative-type), which is

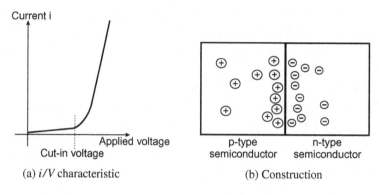

(a) *i/V* characteristic (b) Construction

Figure 4.8 The semiconductor diode

made to have more electrons than normal giving a net negative charge. The surfeit of electrons is introduced by a process called *doping*. Doping can also cause a deficit of electrons and hence a surfeit of *holes* (essentially the hole left by the absent electron), which gives a net positive charge as in a (doped) *p-type semiconductor*. If we join these two types of semiconductor we now have a pair of adjacent materials, one with a net negative charge, the other with a net positive charge; this is called a *pn junction*. Even without any externally applied voltage, electrons in the n-type region will be attracted to the p-type region. There will also be a thermal effect causing electrons to jump the other way. With no externally applied voltage the current generated by thermal effects will balance the current generated by charge effects. When the diode is forward biased, the potential of the p-type region is raised, and so more electrons will migrate to the p-type region causing current to flow. When the potential between the p-type and n-type regions is below the cut-in voltage, it is insufficient to attract electrons in large numbers and no net current flows through the diode.

The current is actually exponentially related to the forward bias voltage and thus can increase dramatically. There is a limit to this current since diodes can break. Also, the reverse breakdown voltage specifies a reverse bias sufficiently large for the diode to pass current in the opposite direction.

Detailed examination of a diode's performance can be found in texts referenced at the end of the chapter. The function of a diode in this text is as a switching circuit; we want it to switch ON and OFF. A diode can turn ON if forward biased and will turn OFF if reverse biased. This is equivalent to applying a logic '1' and '0' (Figure 4.9). Logic circuits based on diodes have poor performance: the fanout is less than the unity so we need to include gain in the circuit.

| (a) Diode switched on | (b) Diode switched off |

Figure 4.9 The diode as a switching circuit

4.2.2 *Bipolar transistors*

A diode has limited functionality in logic circuits and the *transistor* plays a more important role, primarily because it is a three-terminal device and there are two forms:

(i) *Bipolar*: conduction uses both holes and electrons (as in a diode).
(ii) *Unipolar*: conduction uses one type of carrier only, either holes or electrons.

A bipolar transistor can be made by fusing two regions of n-type semiconductor with one region of p-type. The connections to the n-type regions are termed the *emitter* and the *collector*, and the p-type region is called the *base*. This gives an *npn transistor*. It is made from the two types of semiconductor in three regions. Two of the regions are n-type and the other one is p-type. The arrangement is shown in Figure 4.10(a) and this can be viewed as two diodes (of the form of Figure 4.8) which are back to back. Its symbol, Figure 4.10(b), shows a three-terminal device with a diode between the base and emitter.

The diode between the *base-emitter junction* effectively controls the operation of the device. If the base-emitter junction is forward biased then electrons are attracted from the emitter into the base and are swept into the collector. Forward biasing the base-emitter junction then causes a collector current to flow through the transistor. If the base-emitter junction potential is below a threshold value then there will be a small current from base

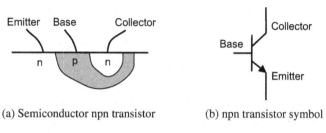

(a) Semiconductor npn transistor (b) npn transistor symbol

Figure 4.10 Transistors

to emitter due to charge effects (electrons attracted into the p-type) and a small current from emitter to base (from electrons migrating due to thermal effects). If the base-emitter junction is not forward biased then there will be no net current since the thermal effects will balance the charge effects and there will then be no collector current. We can use the base-emitter voltage to switch the collector current on or off. When we introduce a power supply for the collector current, together with a resistor (essentially to limit the collector current), we obtain a circuit that can be used as a switch (shown in Figure 4.11 where the collector, emitter, and base have the abbreviated symbols C, E, and B, respectively).

Note that the two pn junctions imply that the transistor can be modelled as two diodes. This is in fact the basis of the Ebers–Moll model of a transistor.

(a) Transistor inverter circuit (b) Transfer characteristic

Figure 4.11 Transistor inverter

The collector current will vary with the base-emitter voltage, which in turn causes the collector-emitter voltage to vary. As we forward bias the base-emitter junction the collector current increases and so the collector voltage falls. When the transistor is turned OFF there is no collector current, so the transistor is OFF, and the collector is connected to the positive voltage supply, so V_{CE} is HIGH. If the base-emitter voltage, V_{BE}, is above a threshold value then the increase in collector current implies a reduction in V_{CE}, which reduces until V_{CE} is LOW. The collector-emitter voltage is the reduction from the power supply voltage by the voltage drop across the load resistor.

$$V_{CE} = V_{CC} - i_C R_L$$

(This is the voltage drop discussed when considering fan-out earlier in Section 4.1.2.) An increase in the collector current i_C implies a voltage drop across the load resistor R_L, causing V_{CE} to be lower than the power supply V_{CC}. There are three named regions associated with the transistor's behaviour that are marked on the transfer characteristic, which shows the change in V_{CE} with V_{BE}. These are:

(i) The **off** region where the collector current is zero.
(ii) The **active** or **linear** region where the collector voltage varies rapidly with changes in V_{BE}.
(iii) The **on** region where the collector current is large.

In the *linear* region the slope of the characteristic implies that a small change in V_{BE} will cause a large change in V_{CE}. This is the performance of an amplifier and the transistor is operated in this region in linear circuits (there is a straight line approximation in this region of the characteristic). In logic circuits we will be concerned mainly with two states, the ON and the OFF state. This circuit operates as a logic inverter since when V_{BE} is HIGH ($V_{BE} = $ '1') then V_{CE} is LOW ($V_{CE} = $ '0'). Conversely, the output is HIGH, '1', when the input, V_{BE}, is LOW and the transistor is switched OFF.

Circuits based on transistors led to transistor–transistor logic (TTL), which dominated logic circuit design at one time with the 74 series (as described earlier in Section 2.4). The technology had reasonable

performance factors, except one. And that one was power consumption. So when devices arrived that had the same speed performance and with much lower power consumption, they became the natural choice for a designer. These devices are based on a different type of transistor.

4.2.3 *Field-effect transistors*

There are also *field-effect transistors*, FETs, which have one type of charge carrier controlled by the doping process, either electrons or holes, but not both (as in a bipolar transistor). The technology used to construct these transistors is *metal–oxide–semiconductor*, MOS, and the transistors are often known as MOSFETs. Field-effect transistors have three terminals but the names on the connections differ from bipolar transistor technology: here we have the drain (replacing the collector), the source (replacing the emitter), and the gate (replacing the base). The aim of the device is to control the current flowing between the drain and the source. This is achieved by changing the voltage applied to the gate. The gate is insulated from the p-type substrate, as shown in Figure 4.12, and so acts as a capacitor affecting the charge. When no voltage is applied, the device is switched off and no conduction occurs between the two regions of n-type material (the drain and the source). When voltage is applied between the gate and the source, as shown in Figure 4.12(b), electrons are attracted into a channel which forms between the drain and the source (the gate/substrate acts as a capacitor and the substrate is the lower plate of the capacitor). This increases the channel conductivity, so current can flow from drain to source, and the device is switched on.

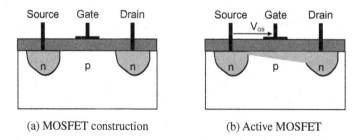

(a) MOSFET construction　　　　　(b) Active MOSFET

Figure 4.12 Basic MOSFET

Figure 4.12 is a simplified picture of a more complicated position. This is an enhancement-mode transistor and suffices here. Again, there are three types of activity:

(i) The **cut-off** region where the transistor is off and the drain current is zero.
(ii) The **linear** or **ohmic** region where the drain voltage varies rapidly with changes in V_{GS}.
(iii) The **saturation** mode where the channel and the drain current are large.

There are two types of transistor consistent with the two types of carrier. One type uses (a channel made from) holes and is positive MOS (PMOS); the other type uses electrons and is negative MOS (NMOS). Essentially, the transistor in Figure 4.12 is an NMOS transistor with its n channel; conversely, if the substrate is n-type and the two connections are p-type then the channel is p-type (and V_{GS} is negative) giving a PMOS transistor. The symbols are shown in Figure 4.13.

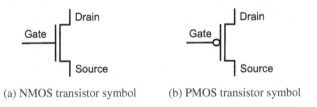

(a) NMOS transistor symbol (b) PMOS transistor symbol

Figure 4.13 MOSFETs

There are actually many different types of FETs. We are using FETs here within logic technology and it is easiest to think of FETs as **voltage-controlled switches (resistors)**. For both FET transistors the *drain source voltage*, V_{DS}, is controlled by the *gate source voltage*, V_{GS}. An NMOS transistor switches ON (the *drain current*, i_D, is positive) for positive V_{GS}. If V_{GS} is less than the threshold then the transistor is OFF (i_D is zero). Conversely, a PMOS transistor is ON when V_{GS} is less than a threshold value (and i_D then increases), when V_{GS} is greater than the threshold the transistor is OFF and i_D is zero. This is why an inverter symbol is included within the PMOS transistor symbol.

> *A bipolar transistor can be thought of as a current-controlled current source (since i_C can depend mainly on the base current i_B when V_{CE} is sufficiently high) whereas an FET is a voltage controlled current source (since i_D can depend on V_{GS} when V_{DS} is sufficiently large).*

A similar circuit can be developed as with the bipolar transistor to give an inverter; here the inverter is based on an NMOS transistor (Figure 4.14).

(a) Circuit (b) Transfer characteristic

Figure 4.14 Basis of NMOS inverter

The power supply is called V_{DD} in MOS technology. Again the circuit is that of an inverter since if an input voltage is applied between the gate and source then a HIGH input switches the transistor ON and the output is LOW, whereas a LOW input (one lower than the threshold to switch on the NMOS FET) switches the transistor OFF and the output is HIGH, connected to V_{DD}. As a logic technology, MOSFETs have major advantages. When switched ON the impedance is low (of the order of 100 Ohms), whereas when switched OFF the impedance is exceedingly large (of the order of gigaOhms [GΩ]). The gate current is almost non-existent, except during switching, because the gate is insulated from the source/drain. This leads to a high fan-out capability, as MOSFETs are currently the closest possible approximation to an ideal electronic switch.

Since resistors need a lot of space in VLSI (very-large-scale integration) circuits (they need plenty of 'real estate', but that sounds a bit too close to domestic housing), resistors can be made using transistors. We then have the circuit using a load resistor replaced with an NMOS transistor, shown in Figure 4.15. The NMOS load transistor is permanently switched on by connecting its gate to the positive power supply. When both transistors are ON then the circuit is equivalent to a potential divider. If both transistors were the same, the output voltage for a '0' would be $V_{DD}/2$. To force V_{OL} to be away from the forbidden zone, we use a process called **ratioing**, which changes the transistors' relative size and thus their resistance, ensuring $V_{OL} \ll V_{OH}$.

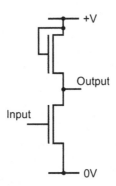

Figure 4.15 NMOS inverter

This gives the circuit for an *NMOS inverter*. This inverter has two states. The first state is where the input is '0' and the lower NMOS transistor is OFF. When it is OFF, it has high impedance and draws no current, and so disappears from the circuit (Figure 4.16[a]). The second state is when the input is '1' and the lower NMOS transistor is ON. Since its impedance is much lower than the load transistor then effectively only the lower transistor remains in the circuit. The output is then connected to ground through the switched ON lower NMOS transistor, shown in Figure 4.16(b).

The circuit for a *PMOS inverter*, which uses PMOS transistors, is shown in Figure 4.17. The circuit using an inverter is the basis in Figure 4.17(a) and the one using just transistors where the resistor is replaced with a permanently switched-on PMOS transistor is shown in Figure 4.17(b).

(a) Input = '0'; Output = '1' (b) Input = '1'; Output = '0'

Figure 4.16 Two states of an NMOS inverter

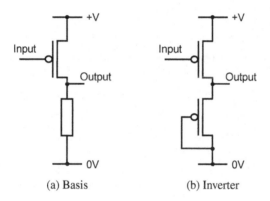

(a) Basis (b) Inverter

Figure 4.17 PMOS inverter

Positive MOS depends on holes as the majority carrier; holes are **slower** due to their decreased mobility when compared with that of electrons. This leads to PMOS being larger than NMOS. Since the smaller size implies smaller capacitances and hence faster circuits, NMOS dominated early MOS technology.

The high resistance of a MOSFET input can introduce susceptibility to *static damage*. We can generate a potential, via static electricity, in excess of 5 kV just by walking across a carpet. When we touch a chip input, the resistance of the air gap between the chip input and our finger is about 1 GΩ. If the chip input circuit has small impedance compared with 1 GΩ, then the potential will be dropped across the air gap and we see a spark. If the impedance of the chip input is large compared with the air gap, the, say, 9 GΩ impedance of a MOSFET is clearly large compared with 1 GΩ, then the potential will be dropped across the chip input circuit. For a MOSFET

this is actually larger than an internal *breakdown* voltage, which specifies a maximum voltage that can be applied across parts of the device. If we exceed it we will damage the device. This is why Complementary MOS (CMOS) devices are known as *static sensitive* and, even though modern designs have protection built into circuits (diodes on the inputs), it is sometimes prudent to connect yourself to 0 V (or any old gas or water pipe will do, anything that goes to ground) to avoid static damage.

4.3 Complementary Metal–Oxide–Semiconductor (CMOS)

Complementary MOS evolved from combining NMOS and PMOS transistors. The difficulties associated with ratioing motivated development which centred on using both NMOS and PMOS FETs. Since these switch ON for inverted inputs, their combination is called complementary MOS (CMOS).

A CMOS inverter has one transistor that switches the output HIGH for a LOW input. This requires a PMOS transistor connecting the output to the positive power supply. A LOW output is caused by a HIGH input, which requires a NMOS transistor connecting the output to the negative power supply.

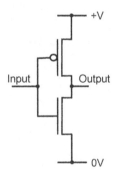

Figure 4.18 CMOS inverter

The equivalent circuit for a CMOS inverter is a pair of switches on which the input acts in a complementary way. The CMOS inverter is at present the closest possible approximation to the ideal inverter, as shown earlier in Figure 4.1.

By comparing the CMOS inverter with the ideal inverter in Figure 4.1, the FETs just replace the switches. For logic gates (such as AND and OR), CMOS circuits actually combine the output stage within the switching function as in Figure 4.19, rather than use a combinational logic function (Figure 4.2).

The operation of the CMOS inverter can be described by a table showing which transistors are OFF or ON in each logic state:

Input	NMOS	PMOS	Output
'0'	OFF	ON	'1'
'1'	ON	OFF	'0'

A two-input CMOS NAND gate is designed to give a LOW output when both inputs are HIGH. This can be achieved by two NMOS transistors in series — N1 and N2 — connecting the output LOW. Conversely, if either or both inputs are LOW then the output should be connected HIGH. This can be achieved by a pair of PMOS transistors, P1 and P2, which are connected in parallel to the positive voltage supply. The combination of these circuits gives the 'push–pull' nature of the output circuit: the active device connections to the power rails give rise to faster switching.

Figure 4.19 A two-input CMOS NAND gate

The activity of the device can be described in terms of which transistors are ON or OFF according to the values of the inputs. When OFF, the transistors are open circuit, when on, they conduct, leading to the output to be connected either to + V ('1') or to 0 V ('0'), as in Figure 4.20.

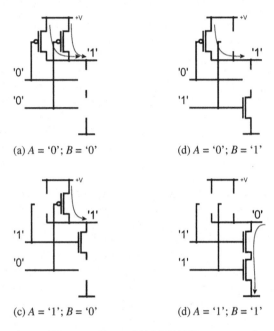

(a) A = '0'; B = '0' (d) A = '0'; B = '1'

(c) A = '1'; B = '0' (d) A = '1'; B = '1'

Figure 4.20 A two-input CMOS NAND gate operation

It can be seen that the parallel PMOS transistors serve to connect the output HIGH if either input is LOW, and the nature of current flow is shown by the dotted lines. The parallel PMOS transistors are actually an OR structure; if input 1 or input 2 is LOW then the output is HIGH. This is an illustration of De Morgan's law:

$$\overline{A \cdot B} = \overline{A} + \overline{B}$$

The NMOS transistors serve to pull the output low when they are both turned ON. This is an AND structure. If input 1 and input 2 are HIGH then the output is LOW. The equivalent circuits for a NAND gate sourcing HIGH and LOW outputs (shown in Figure 4.20) concern only the transistors which are switched ON. The large impedance of a MOSFET which is OFF implies

that it effectively disappears from the circuit. An alternative description of the circuit is a functional table:

Input A	Input B	P1	P2	N1	N2	Output
'0'	'0'	ON	ON	OFF	OFF	'1'
'0'	'1'	ON	OFF	OFF	ON	'1'
'1'	'0'	OFF	ON	ON	OFF	'1'
'1'	'1'	OFF	OFF	ON	ON	'0'

The CMOS two-input NAND gate has two circuits, one of which provides a NAND gate for LOW inputs (the PMOS transistors), and the other is a NAND gate for HIGH inputs (the NMOS transistors). This is reflected in other CMOS gates. The two-input CMOS NOR gate (Figure 4.21) provides an output which is LOW when either input is HIGH and an output which is HIGH when both inputs are LOW. The parallel OR structure is used to give a LOW output. The series AND structure is used to provide a HIGH output when both inputs are LOW.

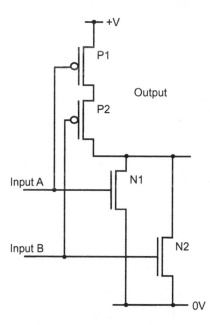

Figure 4.21 A two-input CMOS NOR gate

The activity of the NOR gate can again be described in terms of which transistors are ON or OFF according to the values of the inputs:

Input A	Input B	P1	P2	N1	N2	Output
'0'	'0'	ON	ON	OFF	OFF	'1'
'0'	'1'	ON	OFF	OFF	ON	'0'
'1'	'0'	OFF	ON	ON	OFF	'0'
'1'	'1'	OFF	OFF	ON	ON	'0'

The main advantage of MOS technology is that it has very low power consumption. This is when the device is not switching and is termed the *static power dissipation*. That the static power is low is intimately related to the closeness of FETs to the ideal gate. The large OFF resistance implies very low power consumption. Disadvantages of MOS technology include a poor drive capability for large capacitative loads (the current sourced by MOSFETs can be less than that for bipolar transistors). Also, the power consumption increases with switching speed. The power consumption is actually related to

$$\text{power consumption} \propto f \times V^2_{supply} \times C$$

where f is the switching frequency, V^2_{supply} is the positive voltage supply, and C is the capacitance driven. It can be observed that the power supply plays a major role in CMOS performance characteristics. Both the fan-out of CMOS circuits and the noise margins are good.

> *One advantage of CMOS is that the power supply can range between 1.5 V and 15 V, and this can be used by designers to optimise performance. Note that low power supply logic will continue to be very important in the future.*

The major advantages of CMOS are that current versions can achieve the moderate speeds required for general purpose logic with lower power consumption than other logic technologies. For these reasons CMOS now dominates much of integrated circuit design. The early CMOS *logic family*, which provided a range of logic functions, was the 4000 series introduced

in 1972. This offered very low power consumption but with very low speed operation (about 80 ns propagation delay per gate). This family offered ranges of gates from simple combinational gates to more complex circuits. Though slow, the 4000 series dominated early implementations where power consumption was critical. Developments in CMOS manufacturing technology have led to the present position where CMOS now dominates implementation; this was achieved by supplanting a famous and popular logic technology that used bipolar transistors: TTL logic.

4.4 Concluding Comments and Further Reading

This chapter has described what is inside integrated circuits from the viewpoint of the major logic technology CMOS. As a designer it is often necessary to be aware of the properties of the circuits you are using to implement designs. You will be designing circuits to meet a wide-ranging specification. Some aspects of the specification are often more critical than others and these can dominate the logic technology chosen for implementation. Designers use less prefabricated-function circuits now (those available from logic families), but you might resort to them from time to time. Though most designers use programmable logic implementations, these naturally depend on a logic technology for implementation. There are other logic technologies — some are in a development stage and some are now obsolete, while others have not penetrated the market much. Diode logic is archaic and TTL is very dated too (though not according to some textbooks). Amongst those that never developed is BiCMOS, which aimed to combine the advantages of the speed of bipolar logic with the low power consumption of CMOS. Failure to penetrate the market might imply that the technology did not live up to its early expectations, but there are also commercial considerations which are beyond the scope of this text.

For an introduction to semiconductor physics, try Parker (2004) or Millman and Grabel (1988), which gives a lucid account of microelectronic technology. There are a number of reasonably priced texts which include digital electronic circuits within a complete study of electronics, see, for example, Sedra and Smith (2010). There are books dedicated to particular technologies, such as Elmasry (1983), Weste and Harris (2010), and Alvarez (1989). There again, the CMOS ones are of most interest now, and Weste

and Harris's book (2010) is well known for its excellent coverage. At the time of writing, the world's fastest transistor came from the University of Southampton (Kham *et al.*, 2006) (110 GHz!) though there is naturally considerable interest, and continuing advance, in this topic.

Manufacturers provide databooks on the circuits that they produce, as well as handbooks on how to use them. There are many manufacturers and many vendors (e.g. RS Components and Farnell) whose websites provide datasheets on the products that they supply. We're finished now with the main building blocks for introductory logic design. The real meat (or a tasty quiche) is logic that incorporates memory: for that we need combinational design and the circuits that implement it. Onwards to sequential logic we go.

4.5 Questions

Solutions in Appendix 1.

Q4.1 Develop a CMOS circuit to implement a two-input (positive logic) AND function.

Q4.2 Deploy a CMOS NAND gate as an inverter. Identify which transistors are redundant, and why.

Q4.3 An NMOS NAND gate is made from two transistors, one with ON resistance 200 ohms and the other with an ON resistance that can be controlled. The OFF resistance for both transistors is in the order of $G\Omega$. What ratio of ON resistance must be achieved so that a LOW output approaches $V_{DD}/3$?

Q4.4 Design a CMOS (positive logic) circuit to implement the function

$$f = \overline{A \cdot B + C \cdot D}$$

Q4.5 For a bit of history (play 'Leader of the Pack' by The Shangri-Las when you do this), make a two-input OR gate from two diodes. Now make an AND gate.

Q4.6 For a bit more history (and changing the music to 'Stairway to Heaven' by Led Zeppelin), develop the transistor-based inverter into

a NAND gate. (Hint: there can be more than one emitter on a silicon transistor, since this just requires an extra connection to be made to the semiconductor substrate.)

Q4.7 Play 'Fields of Athenry' by Dropkick Murphys to return you to the third millennium.

Chapter 5

Bistables

5.1 Sequential Logic Concepts

We will now consider logic that includes time and systems that act in a *sequence* of states, unlike the previous systems where the output was a combination of input values. Sequential systems find wide application, such as to implement a sequence of instructions in a computer, or to implement systems with a sequence of events in an industrial product control system. This implies that (see Figure 5.1):

(i) The systems have a **synchronising signal**, called a *clock*, which is similar to a clock in everyday life.
(ii) The systems have to remember where they are in the **sequence**; this requires *memory*.

Sequential logic concerns binary systems that go through a sequence of *states* where the state is the condition at a particular time. Progression between states is synchronised to a clock signal. The clock signal indicates when to move from one state to the next and is usually a square wave signal.

In practice there are usually many states with a more complex flow between them. The quality control system in Figure 2.1 could not actually be achieved by combinational logic alone, since the movement of the production line implies a sequence of operations. That implies a need for memory to remember that state of the system. Here, the states change when the clock is '1'. This is *level-triggered logic* since we are synchronising to a **level** of the clock. When the clock is HIGH we can **change** state; when

101

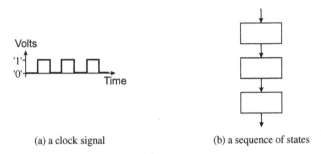

(a) a clock signal (b) a sequence of states

Figure 5.1 Sequencers and sequences

the clock is LOW we **remember** the state. We now have a sequencer, the clock, and devices that have **memory**. These devices are called *bistables*.

5.2 Basic Bistables

5.2.1 *Basis*

A bistable is defined as a device that has two stable states (as implied by its name). In its basic form it comprises a pair of inverters (labelling the outputs Q and \overline{Q} is a convention for bistables since they are usually complementary and one is the inverse of the other) (Figure 5.2). The core of the system is **feedback**: outputs are fed back to become inputs. This is a combinational circuit that uses outputs as inputs to itself.

Figure 5.2 A basis of a bistable

This bistable is perfectly stable in either of two states:

(i) When Q = '1' and \overline{Q} = '0'; and (ii) when Q = '0' and \overline{Q} = '1'.

It can enter either of these states when power is applied. The state in which it actually ends up depends on the propagation delays of the inverters (whichever of the inverters responds fastest determines the final state of the bistable). The final state is actually controlled by feedback since the

output of each inverter forms the input to the other. This feedback provides the memory associated with the device. The two memory states are called *set* and *reset*. A bistable is:

(i) **set** when $Q = $ '1' and $\overline{Q} = $ '0'; and

(ii) **reset** when $Q = $ '0' and $\overline{Q} = $ '1'.

Since we cannot control the memory of this device, we replace the inverters with gates that accept more than one input. We shall replace the inverters with NAND gates and retain the feedback or *steering network*, since this provides the memory by feeding back the current state to the inputs.

5.2.2 *The* $\overline{R}\,\overline{S}$ *bistable*

The $\overline{R}\,\overline{S}$ *bistable* is formed from two cross-coupled NAND gates (Figure 5.3).

Input A — Output Q

Input B — Output \overline{Q}

Figure 5.3 $\overline{R}\,\overline{S}$ bistable

This looks like a combinational logic system, but let us consider what happens when the input A is a '0' and is then changed to a '1' while input B remains unchanged at '1'. These are labelled as input $A = 0/1$ ('0' changing to a '1') and input $B = 1/1$ ('1' remaining at '1'). Noting that the output of a NAND gate is '1' if any input is a '0', the initial state of the bistable will be $Q = $ '1' (since input $A = $ '0') and $\overline{Q} = $ '0' (since \overline{Q} is the output of a NAND gate with inputs $Q = $ '1' and $B = $ '1'). When input A changes from '0' to '1' then $Q = $ '1' since $\overline{Q} = $ '0'; also, $\overline{Q} = $ '0' since $Q = $ '1'. The bistable does not change state (the outputs remain unaffected by the change in input A), and can be interpreted to have memory since the bistable remembers that it was set. This memory is provided by the steering network feeding the outputs back to become inputs, which makes \overline{Q} force Q to remain at '1'.

Similarly, if input A is held at '1' and input B changes from a '0' to '1' ($B = 0/1$), then the outputs will initially be $\overline{Q} =$ '1' (since input $B =$ '0') and $Q =$ '0'. The outputs again remain constant and do not change when B changes since $Q =$ '0' is fed back to keep $\overline{Q} =$ '1'. The memory function associated with changing both inputs HIGH can be summarised as:

Inputs		Outputs	
A	B	Q	\overline{Q}
0/1	1/1	1/1	0/0
1/1	0/1	0/0	1/1

This is termed *latching* — the $\overline{R}\,\overline{S}$ bistable latches, or remembers, the outputs when both inputs are HIGH. This is the property of memory and it is very important. In combinational logic the output depends on the current set of inputs. In sequential logic, the property of memory implies that the current output depends not only on the values of the input but also on the previous output. A summary of the bistable's function (with a lighter output indicating '1' and a darker output indicating '0') is shown in Figure 5.4.

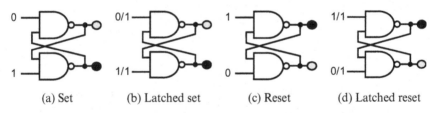

 (a) Set (b) Latched set (c) Reset (d) Latched reset

Figure 5.4 Setting, resetting, and latching the $\overline{R}\,\overline{S}$ bistable

Bistables are intended to have **two** stable states:

(i) **Set** when $Q =$ '1', $\overline{Q} =$ '0'; and (ii) **reset** when $Q =$ '0', $\overline{Q} =$ '1'.

So this bistable latches, or remembers, whether it was set or reset. Input $A =$ '0' sets the device since it forces the output Q to be '1'; it is then usually called \overline{S} since a LOW value sets the device. Input $B =$ '0' resets the device since it forces \overline{Q} to be '1' and thus Q to be '0'; it is then usually called \overline{R} since a LOW value resets the device. So input A in Figure 5.3 is \overline{S} and input B is \overline{R}.

Note that you cannot form practical memory elements
without inversion and they cannot be made from AND or
OR gates alone.

The operation of the bistable is described by a *characteristic table*
which defines the new state of the bistable Q_{n+1} and \overline{Q}_{n+1} as either set,
reset, or latched (when the new state of the bistable Q_{n+1} latches the
previous state Q_n; when latching $Q_{n+1} = Q_n$, and $\overline{Q}_{n+1} = \overline{Q}_n$).

\overline{S}	\overline{R}	Q_{n+1}	\overline{Q}_{n+1}	
0	0	1	1	Invalid
0	1	1	0	Set
1	0	0	1	Reset
1	1	Q_n	\overline{Q}_n	Latch

When $\overline{S} = \overline{R} = 0$ the outputs are *invalid* since they are both '1' and the
bistable appears to be set and reset at the same time. This is sometimes called
unstable (it is manifestly not unstable since both outputs are fixed HIGH)
since if $\overline{S} = \overline{R} = 0$ and we then change to $\overline{S} = \overline{R} = 1$ the output state will
depend on propagation delays at each gate (akin to the basic bistable formed
from inverters, Figure 5.1). For $\overline{S}, \overline{R} = 0, 1$ and $\overline{S}, \overline{R} = 1, 0$ the bistable is
set and reset, respectively. In this state the device acts as a combinational
logic system; the bistable only exhibits memory when $\overline{S}, \overline{R}$ changes
to 1,1.

An alternative formulation of the $\overline{R}\,\overline{S}$ bistable is to replace NAND with
NOR, giving an RS bistable (Figure 5.5).

R	S	Q_{n+1}	\overline{Q}_{n+1}	
0	0	Q_n	\overline{Q}_n	Latch
0	1	1	0	Set
1	0	0	1	Reset
1	1	0	0	Invalid

(a) *RS* Bistable (b) Characteristic table

Figure 5.5 *RS* Bistable and operation

The set and reset states are entered when the inputs S and R are '1' (and the other input is '0'), respectively. The outputs latch for $S = R =$ '0' and are invalid when both inputs are HIGH.

5.2.3 *Switch debouncer — an $\overline{R}\,\overline{S}$ bistable in action*

This is a commonly cited application of a basic bistable. It centres on using a bistable to remove the effects of bouncing in a mechanical switch. Ideally, when a mechanical switch changes from $+5\,V$ to $0\,V$ the voltage should change with it. Unfortunately, mechanical switches bounce constantly and the voltage will change from $+5\,V$ to $0\,V$ and back again several times. This happens on all mechanical switches, even domestic light switches, but these effects are so fast that they are imperceptible to the human eye. They can, however, have unfortunate consequences when used in sequential logic systems. This bouncing is illustrated by the circuit in Figure 5.6, where the output is initially high when the switch is open. When the switch first makes contact the output changes to $0\,V$, but the output returns to high, $+V$, when the switch bounces and contact is broken. When the switch settles in position, in firm contact, the output is low and is $0\,V$. What we believe happens is the ideal performance (Figure 5.6[b]) where the output changes quickly, and only once. In reality the operation is more like Figure 5.6(c) where bouncing occurs.

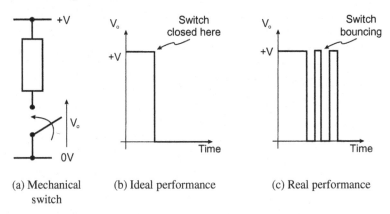

(a) Mechanical (b) Ideal performance (c) Real performance
switch

Figure 5.6 Performance of a mechanical switch

The $\overline{R}\,\overline{S}$ bistable can be used to *debounce* a single-pole double-throw switch, to remove the extra transitions in the output voltage resulting from mechanical bounce. The circuit and its signals are shown in Figure 5.7.

Figure 5.7 Switch debouncer in action

When the switch is thrown from point B to point A at time t it bounces on point A. This means that input B changes from '0' to a '1' when the switch ceases to make contact with B since point B is now unconnected except for the *pull-up* resistor pulling input B HIGH. Conversely, input A is changing between a '1' and a '0' (when the switch connects it is '0' and when the switch bounces away it is '1'). However, the memory function of bistables (via the pull-up resistors) ensures that outputs change only once when the switch position is changed. This is because the inputs A and B both become '1' when the switch ceases to make contact. The bistable outputs track the first change in the switch position (in this case changing the bistable from reset to set), and when the switch bounces this (set) state is latched. So the circuit uses the latch state of the bistable to remember that the switch has been changed, thus debouncing the signals.

This type of switch is known as a single-pole, double-throw switch. The single pole is the earth connection and the (double) throw is the switch connections to point A and point B. Push-button switches (single pole, single way, momentary action) as used in the basic logic circuits earlier, require a different debouncing circuit since there is no common pole for reference for the bistable.

5.2.4 D-*type bistable*

Returning to the basic $\overline{R}\,\overline{S}$ bistable, the main difficulty is the invalid state where both outputs are the same (the device is set and reset at the same time). One solution is to use a single input and invert it between the *R* and *S* inputs. This is the *D-type bistable* (Figure 5.8).

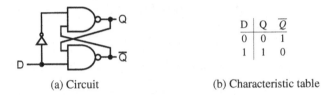

D	Q	\overline{Q}
0	0	1
1	1	0

(a) Circuit (b) Characteristic table

Figure 5.8 Basic *D*-type bistable

The characteristic table for a *D*-type bistable (Figure 5.8[b]) shows that we have avoided the invalid state. However, the basic *D*-type is functionally useless since it merely **delays** the input (an equivalent circuit is that \overline{Q} is derived from an inverter, and *Q* by connection to *D*) and its use comes clearer next.

5.2.5 *Level-triggered* D-*type*

The basic *D*-type becomes useful when a clock is added to give a *level-triggered D-type* (Figure 5.9). The function of this bistable is to accept data when the clock (CLK) is *enabled* or *asserted* at '1' and to latch the data when the clock is *disenabled* or *disasserted*, at '0'. The clock is conventionally gated using NAND gates to allow the data input to affect the bistable output only when the clock is '1'. These NAND gates are coupled to gate the inputs into an $\overline{R}\,\overline{S}$ bistable.

Figure 5.9 Level-triggered *D*-type bistable

When CLK = '0' the previous value of the outputs is remembered, since with the clock at '0' the outputs of both NAND gates are '1' and the latch

state of the $\overline{R}\,\overline{S}$ bistable is being used. When CLK = '1' the input can change the outputs.

D	CLK	Q_{n+1}	\overline{Q}_{n+1}	
X	0	Q_n	\overline{Q}_n	Latch
0	1	0	1	Reset
1	1	1	0	Set

We now have a *one-bit memory* (it latches one bit when the clock is asserted). In order to remember the bit of information, the data must first be set up (the bit to be remembered), then the device is clocked. The bit will then be latched until the following clock cycle. Since the outputs change with the inputs when the clock is asserted (HIGH) it is also called a *transparent D-type latch*.

5.2.6 *Level-triggered* **RS** *bistable (transparent* **RS** *latch)*

A clock input can similarly be connected into an $\overline{R}\,\overline{S}$ bistable to give a *level-triggered RS bistable* (Figure 5.10). We now have two controlling inputs and one clock input, CLK.

Figure 5.10 Level-triggered *RS* bistable and its operation

Its characteristic table is

S	R	CLK	Q_{n+1}	\overline{Q}_{n+1}	
X	X	0	Q_n	\overline{Q}_n	Latch
0	0	1	Q_n	\overline{Q}_n	Latch
0	1	1	0	1	Reset
1	0	1	1	0	Set
1	1	1	1	1	Invalid

This provides the functionality of the $R\,S$ bistable only when the clock is asserted. As with the level-triggered D-type the outputs will change with any change in the inputs for CLK = '1', and so it is also a **transparent latch**. Note that $S =$ '1' and $R =$ '0' sets the bistable (for a clock CLK = '1'), $R =$ '1' and $S =$ '0' resets it. Note also there is still an **invalid** state (for $R = S =$ '1', $Q = \overline{Q} =$ '1').

5.2.7 *Level-triggered* JK *bistable*

The *level-triggered JK bistable* (Figure 5.11) avoids the invalid state by feeding the outputs back twice, i.e. using a double steering network. The invalid state is replaced with a *toggle* state where the outputs flip to become the complement of each other: $Q_{n+1} = \overline{Q}_n$ and $\overline{Q}_{n+1} = Q_n$. All other states are the same as for the $R\,S$ type.

The D in the D-type stands for delay, the RS bistable is well labelled, but what about the JK? Who knows — the origins of the JK appellation are forgotten now. (It's the same for the label Q.)

Figure 5.11 Level-triggered *JK* bistable

J	K	CLK	Q_{n+1}	\overline{Q}_{n+1}	
X	X	0	Q_n	\overline{Q}_n	Latch
0	0	1	Q_n	\overline{Q}_n	Latch
0	1	1	0	1	Reset
1	0	1	1	0	Set
1	1	1	\overline{Q}_n	Q_n	Toggle

If you analyse this circuit you will find that it actually toggles continually when both inputs are HIGH and the clock is HIGH. It then has no

function as a level-triggered device but is included (again) for completeness within the set of basic bistables. It only exists as an edge-triggered device, which comes later in Section 5.5.3. The *JK* bistable is rather archaic now and design centres mostly on use of *D*-type bistables. If you need a *JK*, it is easy to make one from a *D*-type and this will be illustrated in Section 6.2.2.6.

5.3 Clocks and Synchronisation

We have so far considered level triggering where events are synchronised to a HIGH level of the clock. When the clock is LOW, the device is latched and the outputs are static and do not change with the inputs. Synchronism could easily be based on a LOW clock level with devices latched for a HIGH clock. It does not really matter which way round the clock is. A major problem is that the outputs can change all the time the clock is asserted. We usually need a more precise time for synchronism, to synchronise to a much smaller portion of the clock. For this reason we use the edge of the clock, termed *edge triggering*.

synchronising to a HIGH level, states change for CLK.H
synchronising to a LOW level for CLK.L
synchronising to a rising edge
synchronising to a falling edge

Figure 5.12 Synchronising sequential systems

There is a range of symbols which indicate whether the part of the clock used for triggering is an active HIGH level (outputs change with inputs for CLK = '1'), active LOW level (outputs can change for CLK = '0'), positive edge (the outputs assume the value of the inputs just prior to the clock changing from LOW to HIGH), and negative edge (where outputs assume input values just prior to the falling edge of a clock). *Synchronous sequential systems* are those where all sequential devices are driven by the same clock event (Figure 5.12). For many systems this concerns edge triggering, and all elements are triggered at exactly the same time.

These symbols (Table 5.1) are shown both as the symbol used in the characteristic table describing the sequential circuit's operation, and as the bistable symbol. These days, the inputs are depicted on the left,

and the outputs on the right. In older drawing conventions, the clock is depicted as entering from the bottom. The symbols used for bistables in some simulators such as Logisim are not as conventional as these.

Table 5.1 Bistable symbols

Clock Event	Characteristic Table Clock Symbol	Bistable Symbol
Positive Level	'1' or H	D Q / CLK Q̄
Negative Level	'0' or L	D Q / ○CLK Q̄
Positive Edge	↑	D Q / ▷CLK Q̄
Negative Edge	↓	D Q / ○▷CLK Q̄

There are asynchronous circuits too. In asynchronous circuits sequential elements are not all driven by the same clock event. Also, the term 'bistable' is sometimes inferred to mean an edge-triggered bistable and a 'latch' to mean a level-triggered version. The term bistable is now predominantly ubiquitous and refers to either type.

5.4 Master-Slave *D*-type Bistable

The *master-slave bistable* (Figure 5.13) was one of the earliest forms of edge triggering. It uses two *D*-type transparent latches where the clock of one is inverted to provide the clock for the other. The two *D*-types are connected in series and the clock between them is inverted.

Figure 5.13 Master-slave *D*-type bistable

The inversion of the clock between the two stages provides a form of edge triggering since the first *D*-type is enabled when the clock is HIGH and the second *D*-type afterwards, when the clock is LOW. Then the output of the first bistable will follow any change in the input when the clock is HIGH, but will be fixed to their last state (when the clock was HIGH) when the clock returns LOW and the first bistable latches. This value will be latched by the second bistable since its own clock is now HIGH. The phases of the clock signal are as shown in Figure 5.14.

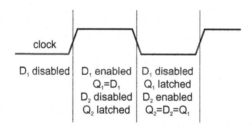

Figure 5.14 Clock phases for master-slave *D*-type bistable

The output Q_1 (and hence the input D_1) **just prior** to the falling clock edge is latched as the output of the device. By inverting the clock between bistables the inputs never directly affect the output hence avoiding the **transparency** problem in the basic latches. The outputs of the second bistable are a latched version of the outputs of the first bistable and are constant. Though the device actually latches data values just prior to the falling edge of the clock, the master-slave bistable is not a true edge-triggered device since it needs both halves of the clock; it is actually *pulse triggered* though some books refer to it as edge triggered.

The name 'master slave' is perhaps the nastiest in digital electronics. Is one the whipping post for the other?

The master-slave bistable's characteristic table is

D	CLK	Q_{n+1}	\overline{Q}_{n+1}	
X	0	Q_n	\overline{Q}_n	Latch
0	⅃	0	1	Reset
1	⅃	1	0	Set

The timing for the master slave bistable is then shown in Figure 5.15 where the shaded areas indicate where Q_1 and Q_2 are latched. Here, we can see that the output of Q_1 is transferred to become the output of Q_2 just after the falling edge of the clock, and thus becomes the bistable's output. As such, the bistable registers the data input existing just prior to the falling edge of the clock (given that a pulse of the clock has occurred).

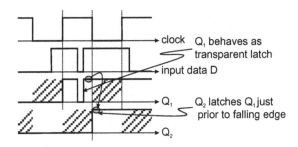

Figure 5.15 Timing within a master-slave bistable

5.5 Edge-Triggered Bistables

5.5.1 *D-type function*

The *positive edge-triggered D-type bistable* is a true edge-triggered device. Input data just prior to a rising clock edge becomes the outputs when the clock changes from '0' to '1'. Input data is then transferred to the outputs at the positive edge of the clock signal. This synchronises change to an extremely small portion of the clock that has particular advantages in practical designs. Its symbol and operation, summarised

using a characteristic table, are:

D	CLK	Q_{n+1}	\overline{Q}_{n+1}	
X	0	Q_n	\overline{Q}_n	Latch
X	1	Q_n	\overline{Q}_n	Latch
0	↑	0	1	Reset
1	↑	1	0	Set

We shall omit the two latch states from now on, since they are implicit in an edge-triggered system. This characteristic table can also be interpreted as:

Present State		Next State
Q_n	D	Q_{n+1}
X	1	1
X	0	0

where the transition between present and next state occurs at **active clock transitions**. This allows us to consider the output of a sequential device to depend on the current value of the output (the current state) and the current value of the inputs. We shall use such concepts later in the analysis and design of sequential systems.

5.5.2 *Edge-triggered D-type circuit*

A circuit for a positive edge-triggered D-type is pretty complex and is given in Figure 5.16. Essentially, operation depends on the propagation delays of the NAND gates. When the clock changes from '0' to '1', the middle set of NAND gates is enabled. The central bistable's state is fed back to gate the input, which is the single D input, and forward to the output stage. The gating stage at the input locks out any further change in the input D since the gating will take effect after the bistable has registered the data. When the clock returns to zero, the lock out is not required since the central bistable will return to a state where both outputs are '1', which means the final bistable is latched and will not change. You can simulate this in Logisim, but correct operation is not guaranteed as it's a Java-based system and the time delay is the operation execution time. This circuit in Figure 5.16

can be optimised — as it in integrated circuit (IC) design — but it is clearer (yes, honestly) as it stands.

Figure 5.16 Positive edge-triggered D-type bistable circuit

5.5.3 *Other edge-triggered bistables*

The positive edge-triggered D-type bistable is the main workhorse in sequential system design. There is naturally a negative edge-triggered version. There is also the *edge-triggered JK bistable*. The *JK* bistable is often made from an edge-triggered D-type (we shall see how later on). The last type of bistable is derived from a *JK* bistable and is used to simplify notation. The *toggle-type bistable* (*T-type*) is one that is either latched or toggled: the outputs remain the same at an active clock edge, or become inverted. This is controlled by a T input, as shown in the following characteristic table.

T	CLK	Q_{n+1}	\overline{Q}_{n+1}	
0	↑	Q_n	\overline{Q}_n	Latch
1	↑	\overline{Q}_n	Q_n	Toggle

This can be implemented using a *JK* bistable where the inputs are connected together. The symbols for the (edge-triggered) *JK T*-type and the *T*-type bistables are shown in Figure 5.17.

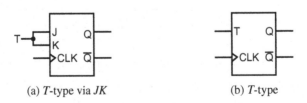

(a) *T*-type via *JK* (b) *T*-type

Figure 5.17 Positive edge-triggered *JK*- and *T*-type bistable symbols

To implement a T-type bistable using a D-type, we need a system which latches when $T = $ '0' and toggles when $T = $ '1'. When it latches the next output Q does not change and when latching it is equivalent to feeding back the Q output to the D input. When it is toggling, Q becomes \overline{Q} so this is equivalent to feeding back the output \overline{Q} to become the input D. The toggle input T controls which of these two conditions are selected, so the function of a T-type bistable is to use the T input as the input to a multiplexer controlling the selection of which output is fed back, as shown in Figure 5.18(a). The full circuit for a T-type constructed using a D-type bistable is then given in Figure 5.18(b).

(a) Using a multiplexer (b) Using logic gates

Figure 5.18 T-type bistable constructed from a D-type

By inspection of the circuit in Figure 5.18(b), we find that the equation defining a D-type constructed from a T-type is

$$D = \overline{T}Q + T\overline{Q}$$

As such, when the T input is '1', the bistable will toggle permanently and the circuit in this state is a bistable with the \overline{Q} output fed back to the D input.

5.6 Timing Considerations

Clocks do not change infinitely fast and so timing is usually referred to a **threshold** point and manufacturers often use a 50% point, halfway between HIGH and LOW. Data must be **set up** before the clock edge and T_{SU} ($T_{SET\ UP}$) defines how long before the clock edge the data must be ready and stable. Data must be **held** constant after the clock edge and T_H (T_{HOLD}) defines for how long data must be held (Figure 5.19).

Digital Electronics: A Primer

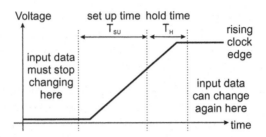

Figure 5.19 Timing considerations for a positive edge-triggered *D*-type bistable

Changing data inputs just after the timing edge can cause the wrong data to be latched, but in many devices the hold time, T_H, is zero. Violating the set-up time T_{SU} may cause **unpredictable behaviour**; referred to as *metastability*. The bistable input data should never change within the window defined by the *set-up* and *hold* times.

Metastability describes a bistable output between its stable states (HIGH or LOW). Any bistable output can reach such a state, and though momentary the fast oscillation can have serious consequences.

Changes in the inputs take some time to propagate through to the outputs and this gives rise to *propagation delay*, which can differ between the output changing from LOW to HIGH and from HIGH to LOW. As in Figure 4.6, the propagation time for an output changing from HIGH to LOW, T_{PHL}, and the propagation delay for an output changing from LOW to HIGH, T_{PLH}, defines how long after the clock edge outputs will change, either from HIGH to LOW or from LOW to HIGH, respectively. This is effectively how long you have to wait to see a result. These times are usually quoted by manufacturers for devices as **worst case** to guarantee a product's performance. Together, these timing parameters then give the maximum clock frequency at which you can drive devices.

The sequence of events for a bistable in a sequential system are:

(i) First the data must be set up.
(ii) After the clock event the data propagates to the outputs.

(iii) Wait for the outputs to propagate through any combinational logic.
(iv) New data is set up ready for the next clock event.

The *minimum clock period* is then the total time for the first three elements of this sequence (the fourth contributes to the following clock cycle). The minimum clock period T_{min} is the reciprocal of the maximum clock frequency, f_{max}, and is given by

$$T_{min} = \frac{1}{f_{max}} = T_{SU} + T_P + T_C$$

where T_P is the larger value of T_{PLH} or T_{PHL}. This is the worst-case value, the longest time to respond, as necessary for calculating the *maximum possible frequency* at which we can issue clock edges and still obtain a reliable output. The combinational logic delay, T_C, implies that the next state of the bistable depends on a combination of logic signals, again this should be a worst-case value — the longest path that any combinational signals affecting the input have to traverse following a clock edge. Since the hold time is invariably less than the propagation time, it does not affect the minimum clock period and therefore does not appear in its calculation.

The circuit in Figure 5.20 will toggle when the enable signal is HIGH and be continuously LOW when enable is LOW.

Figure 5.20 Bistable circuit

The maximum clock frequency for the circuit in Figure 5.20 can be calculated from the timing parameters. For the bistable $T_{PLH} = 0.2$ ns, $T_{PHL} = 0.2$ ns, and $T_{SU} = 0.05$ ns, and the worst-case propagation delay of the AND gate, $T_P(\text{AND})$ is 0.05 ns. The minimum clock period is then

$$T_{min} = T_{PHL} + T_P \text{ (AND)} + T_{SU} = 0.5 \text{ ns}$$

This corresponds to a maximum clock frequency of 2 GHz.

5.7 Asynchronous Inputs

When you first apply power to a bistable it will be set or reset at random according to local conditions; most bistables do not conveniently assume the reset state when you apply power, it needs to be forced. Two *asynchronous inputs* can be used to control the state of a bistable without reference to the clock signal (Figure 5.21). This is why they are called asynchronous, since their action is not synchronised with the clock. The inputs are:

(i) *Preset* \overline{PR} = '0' (active LOW) sets the bistable and sets Q HIGH.

(ii) *Clear* \overline{CLR} = '0' clears it and resets Q LOW.

The last stage of a bistable circuit is an $\overline{R}\,\overline{S}$ bistable and this is where the asynchronous inputs are included.

Figure 5.21 Basis of asynchronous inputs

When bistable outputs are stable (after the propagation time is completed) the last $\overline{R}\,\overline{S}$ stage is then latched. The preset input, \overline{PR}, can be used to set the bistable; the clear input resets it. These inputs should always be complementary when used and their function is usually included within a characteristic table as:

Asynchronous Inputs		Clock	Output
\overline{PR}	\overline{CLR}	CLK	Q
0	1	X	1
1	0	X	0

These inputs should be used with **caution**. If they are not used, connect them to their inactive state, usually to the $+V$ power supply. The way they are included in the bistable is actually more complicated than it appears in Figure 5.21. The preset and clear inputs cause the cleared and set states to

persist. As such, the asynchronous inputs also need to be fed throughout the bistable, to preset and clear the preceding bistable stages, as shown in Figure 5.22. This is because the bistable is made from three interconnected bistables and all of them need to be preset (or cleared). In Figure 5.22, the different types of dotted lines show the connection of the preset and clear signals.

Figure 5.22 Including preset and clear inputs

5.8 Concluding Comments

This chapter has been devoted to the main components of sequential logic, those with memory: bistables. This memory is achieved by using feedback with which the current state of the device can be retained. Of the triggering systems, edge triggering is by far the most important since the more reliable systems are those where change is synchronised to a very short portion of the clock. Of the bistables, the *D*-type is ubiquitous now. Some sequential devices appear to have a *JK* bistable, but closer inspection of the circuit reveals that this is actually constructed using a *D*-type. As such, we now have the major components of sequential design and these will be used throughout the next chapter.

Very little is revealed about the insides of bistables in commercial circuits. As noted, there is actually some optimisation possible of the edge-triggered *D*-type circuit given here, but that is beyond this text. An account of the history of bistables would be an interesting bit of technical history, but time has moved on rather too fast for this. As ever, if you find this material hard to digest, get on a simulator quick. They can be a bit unreliable as they are software and not hardware, but they do let you see how things change.

5.9 Questions

Solutions in Appendix 1.

Q5.1 Implement the switch debouncer circuit again for a single-pole, double-throw switch, but using an $R\,S$ bistable instead of an $\overline{R}\,\overline{S}$ bistable.

Q5.2 Implement a level-triggered D-type bistable using NOR gates instead of NAND gates.

Q5.3 For a D-type bistable, $T_{SU} = 10\,\text{ns}$, $T_H = 2\,\text{ns}$, and $T_{PHL} = T_{PLH} = 13\,\text{ns}$. What is its maximum clock frequency?

Q5.4 Implement a T-type bistable using a D-type and AND and OR gates. For the D-type bistable, $T_{SU} = 10\,\text{ns}$, $T_H = 2\,\text{ns}$, and $T_{PHL} = T_{PLH} = 13\,\text{ns}$. For the AND and OR gates $T_{PHL} = T_{PLH} = 12\,\text{ns}$. Calculate the maximum clock frequency for the T-type bistable you have designed.

Q5.5 The circuit in Figure 5.23 aims to provide an output Q whose frequency is 15 MHz when the clock frequency is 30 MHz.

Figure 5.23 Example circuit

For the D-type bistable, $T_{SU} = 10\,\text{ns}$, $T_H = 2\,\text{ns}$, and $T_{PHL} = T_{PLH} = 13\,\text{ns}$. For the AND gate $T_{PHL} = T_{PLH} = 12\,\text{ns}$. Explain what happens in the clock cycle immediately following a rising clock edge. Will the circuit satisfy its aims?

Q5.6 Design a circuit using a push-button switch that can use the preset and clear inputs to control the state of a bistable.

Chapter 6

Sequential Logic Design

6.1 Synchronous Sequential Systems

6.1.1 *General sequential system*

Asequential system is one that moves through a sequence of states, entering each new state at active clock events. The next state following a clock event is determined by the current position in the sequence and by external inputs. Each state is indicated by the bistables within a system and the bistables will be either set or reset. The bistable outputs give the current state. In order for this to be used to determine the next state, the outputs must be fed back as bistable inputs. These are combined with any external inputs to determine the next state of the system. In a *synchronous sequential system*, all bistables have the same clock; in an *asynchronous sequential system* bistables can change state at different times. Most of this chapter concerns synchronous design, since synchronous design leads to reliable designs. Later there is a small section on asynchronous design, since asynchronous design can be used for a small selection of specialised circuits. A general schema for design is shown in Figure 6.1, using the bus convention from Figure 1.4.

The combinational circuitry, combining the current state or bistable outputs with external or control inputs, is often referred to as the *next state logic*. There are *formal techniques* to design the next state logic. The formal techniques will be described after sequential system software specification and an introductory study of basic sequential devices to highlight how they operate.

123

Figure 6.1 General sequential system

There are a number of versions of Figure 6.1. The inputs can be combined with gated outputs or by using multiplexers. All versions achieve the same aim.

6.1.2 *Synchronous sequential systems and software specification*

The major designs we shall study at this level are synchronous sequential systems, as these dominate sequential logic circuit design. Rather than launch into a full design technique, we shall at first describe synchronous sequential systems using software, as this gives a functional overview into many of the design concepts.

In order to provide a software specification of a sequential system we need to introduce triggering or synchronism into the software. The design of most large-scale sequential circuitry now uses edge-triggered *D*-type bistables for the majority of implementations. For *D*-types, data is transferred to the outputs from a single input. Data is therefore **assigned** to the output and so for sequential system software specification we shall use the assignment operator, =, to denote the clock event when assigning values to bistable, termed *registered* (as explained later in Section 6.3.3) outputs. The software description for an edge-triggered *D*-type is then:

```
IF (data) THEN Q_output=HIGH ELSE Q_output=LOW
```

However, since the registered output follows the input directly it is easier to describe a *D*-type as merely

```
output = data
```

and the output assumes the data input at clock events. In a sequential system the next state could be a function of the current state only, without external inputs. In such systems, the current inputs to the bistables are a fed-back combination of their outputs. An appropriate software description is to use a CASE statement, where the next state is determined from a *look-up table* of current states. The look-up table comprises the condition of each bistable, whether it is set or reset.

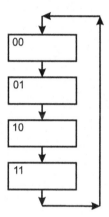

Figure 6.2 A sequence of states

To implement the sequence of states in Figure 6.2, the software description could be (remembering that ";" introduces a comment):

```
CASE (current state)
BEGIN

   0: next_state=state_1
      ;if q1 and q2 are both 0 go to state 1 at next clock edge
   1: next_state=state_2
      ;if q1=0 and q2=1 then go to state 2 at next clock edge
   2: next_state=state_3
      ;if q1=1 and q2=0 then go to state 3 at next clock event
   3: next_state=state_0
      ;if q1=q2=1 then go to state 0 at next active clock
END
```

We need to describe bistable outputs using a *vector* that gives a shorthand way of referring to all bistable values, i.e. to the current state of the device. For two outputs Q_1 and Q_2 we could combine them in a vector called `outputs`, which is of dimension 2 (has two elements). We can then use the first element `outputs[1]` to refer to Q_1, and the second element `outputs[2]` to refer to Q_2. When referring to the state of the device we can refer to `outputs[1..2]` and this can be used either to return the current state of the device (whether the bistables are set or reset) or to assign a new state.

```
CASE (outputs[1..2])
BEGIN
    0: outputs[1..2]=1
       ;if q1 reset and q2 reset, q2 is set at next clock edge
    1: outputs[1..2]=2
       ;if q1 reset and q2 set then set q1 and reset q2
    2: outputs[1..2]=3
       ;if q1 set and q2 reset then set q2 at next clock event
    3: outputs[1..2]=0
       ;if q1 and q2 both set, reset both at next active clock
END
```

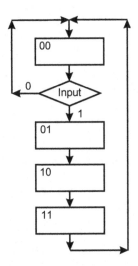

Figure 6.3 Including an external control input

If an external input is included, we need to decide what action to take. We then need to include a decision box within the chart describing the sequence of states and we choose an appropriate branch in the chart according to the value of the tested variable or input. Figure 6.3 shows a chart with a system that has a different sequence of states for different values of the external input. At the first decision box the external control input is tested and, if it is HIGH, the system moves to a state different from that entered when the control input is LOW. The system moves through the system of states only when the external input has been HIGH. If it remains LOW, the system remains in the original state. A software description is then:

```
CASE (outputs[1..2])
BEGIN ;this description is derived from the state chart
    0: IF (input) THEN outputs[1..2]=1
        ELSE outputs[1..2]=0 ;if input=1, go to
                                        ;state 1, else 0
    1: outputs[1..2]=2    ;move to state 2
    2: outputs[1..2]=3    ;move to state 3
    3: outputs[1..2]=0    ;go back to start
        ;always returns to state 0 after state 3
END
```

The IF statement could have been written outside the CASE statement to provide a different software description but one that describes the same function.

6.1.3 *Registers*

A *register* is a set of latches that are connected in parallel, where are all the latches are driven by the same clock, shown in Figure 6.4. Naturally, its function is to **register** or latch data presented in **parallel**. In computer architectures, information is stored as a pattern of bits. This information can be the program that the computer is designed to execute or the data that it is intended to process. We need to latch data as part of a processing cycle and this is achieved using a register. It is then a basic part of any sequential system.

This can use either level-triggered or edge-triggered bistable elements, and edge-triggered elements are shown in Figure 6.4. There is no combinational circuitry associated with this register; in operation we first set up the data, then issue a clock event to transfer the data to the contents of the register, and then hold the data for some time after (according to the timing parameters associated with the device). The register then contains the information that it was designed it to hold. Since it stores a number of bits all presented at the same time, it then has the facility to register *parallel data*.

Since the register latches parallel data, bistable outputs are referred to as registered outputs in software. For a vector of registered outputs of D-type bistables, outputs [1..3], and a vector of inputs, inputs [1..3], a software description assigns data to become outputs following a clock edge and is then simply

```
outputs[1..3]=inputs[1..3] ;register where each output takes
                            ;the value of its associated input
```

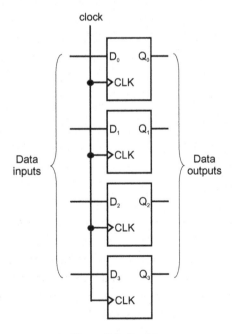

Figure 6.4 Register

6.1.4 *Shift register*

A *shift register* is a device that allows you to move (shift) data within it. The data is shifted from one bistable to the next and its architecture is then a series of bistables with outputs connected to become the input to the next bistable, as in Figure 6.5.

Figure 6.5 Shift register

Assuming that all bistables are cleared before we start (using the asynchronous inputs), then given data which is '0' at the first rising clock edge, '1' at the next, '0' at the third, '1' at the fourth and fifth clock edges, and '0' thereafter, we can depict the data as a time-varying signal, as in Figure 6.6.

Figure 6.6 Data and clock

The arrows on the clock signal signify that the timing charts refer to a positive edge-triggered system. For analysis we shall assume that all bistables are reset with $Q = '0'$ prior to the first clock pulse. At the first clock pulse the input to bistable 0 will be '0' (the input data) and the inputs to all other bistables will be '0' since all bistables are reset. At the first clock edge no bistable will change state since all outputs and the input are '0' and so all the bistables will remain reset. At the second clock pulse the input to bistable 0 will be '1' (the input data), whereas the inputs to all others will still be '0'. After the second clock pulse bistable 0 output Q_0 will be '1' and all others will remain at '0'. At the third clock pulse the input to bistable 0 will be '0' and the input to bistable 1 will be '1' since

Digital Electronics: A Primer

$Q_0 = $ '1'. All other bistable inputs will still be '0'. After the third clock pulse, $Q_0 = $ '0' and $Q_1 = $ '1', and bistable output Q_2 is still '0'. The '1' remembered by bistable 0 at the second clock pulse has then moved, or shifted to, bistable 1 at the third clock pulse. Input data then moves from left to right by one clock pulse at each clock event. This can be summarised in a timing diagram, as shown in Figure 6.7

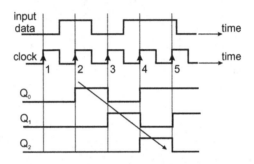

Figure 6.7 Timing diagram for shift register

Note that the shift register accepts input data one bit at a time and it is then said to latch *serial data*. The initial diagram and subsequent analysis has been based on edge-triggered elements. Were you to attempt to build a shift register from level-triggered elements, then problems associated with transparency would arise. Since the outputs of a transparent latch can change whenever the clock is asserted, then the first input data '1' will appear at the output Q_0 and go immediately to the input of bistable 1, appear at Q_1 and pass to Q_2. If level-triggered bistables were used, data would simply fill up the entire register and no functionality would ensue. A shift register **must** use edge-triggered bistable elements. This is shown in the film in Figure 6.8 (sorry, I couldn't resist it — OK, film is rather dated now but we gotta liven this stuff up somehow...) where the data input is that just before the clock rising edge and the bistable outputs are those just after the clock event. The system starts with all inputs and outputs LOW and at the first clock edge the data input is high so the output (of the first bistable) Q_0 becomes HIGH whereas all other bistable outputs remain LOW. At the next clock edge, the data is transferred to become the output Q_1, thus following the timing diagram of Figure 6.7. It carries on like that, shifting the data from left to right.

Figure 6.8 Operation of a shift register

A software description for a vector of registered outputs, outputs[0..2], where outputs[0] $= Q_0$, outputs[1] $= Q_1$, and outputs[2] $= Q_2$. The software description can use the state of each bistable (whether it is SET or RESET), noting that the registered output outputs[0] follows the input data. We are using a binary state notation here (using #b) so that the bit values are clear.

```
;3-bit shift register
CASE (outputs[0..2])
BEGIN     ;in the next state:q1=data, q2=q1, q3=q2
  #b000: IF (data) THEN outputs[0..2]=#b100 ;here data = 1
                   ELSE outputs[0..2]=#b000 ;data = 0
  #b001: IF (data) THEN outputs[0..2]=#b100
                   ELSE outputs[0..2]=#b000
  #b010: IF (data) THEN outputs[0..2]=#b101
                   ELSE outputs[0..2]=#b001
  #b011: IF (data) THEN outputs[0..2]=#b101
                   ELSE outputs[0..2]=#b001
  #b100: IF (data) THEN outputs[0..2]=#b110
                   ELSE outputs[0..2]=#b010
  #b101: IF (data) THEN outputs[0..2]=#b110
                   ELSE outputs[0..2]=#b010
  #b110: IF (data) THEN outputs[0..2]=#b111
                   ELSE outputs[0..2]=#b011
  #b111: IF (data) THEN outputs[0..2]=#b111
                   ELSE outputs[0..2]=#b011
END
```

An alternative software description could use the functionality of the device and is much simpler. Given the same vector of outputs, the description could be:

```
;3-bit shift register
BEGIN
outputs[0]=data
outputs[1]=outputs[0]
outputs[2]=outputs[1]
END
```

This is clearly a much simpler description, but is only achieved by intimate knowledge of the functionality of the shift register and as such can only be achieved for simple systems.

6.1.5 *Bidirectional shift register*

The shift register just presented can only shift data from left to right. This is clearly schematic according to the circuit layout and we could equally have a shift register which shifts the other way, from right to left, as in Figure 6.9.

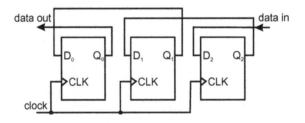

Figure 6.9 Right-to-left shift register

It is often required to have a shift register which shifts information either way, either from left to right or from right to left according to the status of a mode signal. This is a *bidirectional shift register*, shown in Figure 6.10. We then need to choose as a bistable input either data coming from the left or data from the right and a circuit to ensure that the appropriate data is selected. We need the function of a multiplexer, the ability to select either of two inputs according to a mode signal. By attaching a multiplexer

to each input and by driving the multiplexers with a common control signal we then achieve a register with bidirectional functionality, which shifts data left if the mode signal control is HIGH and right when control is LOW.

Figure 6.10 Bidirectional shift register

This sequential system has a state that corresponds to the data stored within the bistables. The next state depends on the current state together with the status of a (external) mode signal. We are then combining bistable outputs with external inputs to determine the sequence of states of the bistable. This accords with the general schema of a sequential system described earlier. Sequential systems rarely fit into a prescribed register pattern, so it is now appropriate to develop a formal technique for sequential system design and synthesis.

```
;Bidirectional 3-bit shift register
BEGIN
IF right_shift
    THEN BEGIN ;accept data from left and shift right
            outputs[0]=L_to_R_data
            outputs[1]=outputs[0]
            outputs[2]=outputs[1]
         END
    ELSE BEGIN ;accept data from right and shift left
            outputs[0]=outputs[1]
            outputs[1]=outputs[2]
            outputs[2]=R_to_L_data
         END
END
```

A bidirectional shift register implements the shift/rotate instructions used in assembly language programming. In that way, it gives a fast way to divide integers by two (shifting right), or to multiply them by two (shifting left). For more details, look at Chapter 7.

6.2 State Machine Design

6.2.1 *The algorithmic state machine (ASM) method*

In order to design general purpose sequential logic systems the design first must be partitioned into a *control algorithm*, which specifies the circuit operation, and an **architecture**, which **implements** the control algorithm. The control algorithm dictates how the circuit operates at an abstract level and may be independent of the hardware solution. When formulated correctly the control algorithm is translated into hardware to implement a design.

A formal method suited to the development of control algorithms for synchronous sequential circuitry is the *algorithmic state machine* (ASM) chart. An ASM chart is superficially a flow chart, although expresses a concept of time intervals in a precise way, whereas a flow chart defines a sequence of events and not their time duration. An algorithmic state machine moves through a sequence of states, where each state is a unique position in a control algorithm, defined by the previous state and some status variables or inputs. For synchronous systems, state times are defined by the clock signal; active clock transitions cause a change from *present state* to *next state*.

The basic symbols for ASM chart notation are those shown in Figure 6.11 and those already used for a state, denoted by a square box with a name attached to it, and a decision box (or conditional branch), where the route taken depends on the value of the status variable tested in the decision box.

Any value written inside the state box is an output generated uncon-ditionally in that state, here any signal allocation written in *italics* denotes an output in that state. The control algorithm may be expressed as an ASM chart using these symbols. The circuit design is the translation of the ASM chart into hardware. As an example, the ASM chart for a two-bit shift

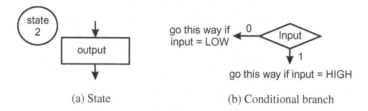

(a) State (b) Conditional branch

Figure 6.11 Basic ASM symbols

register is given by a sequence of states that are followed according to the input data. Each new state then depends on the value of a control input and so each state is followed by a conditional branch. At each clock edge the system moves from one state, through the conditional branch, to a new state that depends on the input data. For a shift register, assuming all bistables start reset, the first state is with all bistables at '0'. If the following input data, at the next clock edge, is '0' the device remains reset. When a '1' is presented this will be latched by Q_A. If it is followed by a '1' the shift register fills with '1's, whereas if it is followed by a '0' the first '1' will be remembered in the second bistable, Q_B, and the '0' in the first, Q_A. The ASM chart is shown in Figure 6.12. It is tempting to think of this in terms of shift-register hardware. That temptation should be resisted for now as we are here considering the flow of information, and covering all possibilities.

There are a number of ways to translate the ASM chart into hardware. It is therefore possible to use a single bistable to remember each state. The position in the ASM chart sequence can then be determined from which bistable is set. This is actually called the *one hot* solution. It is more efficient to *code* a state using a number of bistables. It is possible to **encode** 2^N states using N bistables giving 2^N unique codes, as each bistable is either set or reset in a particular state. This clearly reduces the number of bistables used but the penalty is that this approach can incur more combinational logic (this is inherent in using bistables to code each state). For example, an ASM chart with eight states can be implemented using three bistables; the outputs of the bistables need to be **decoded** to determine the present state. The outputs for each state are written above the box signifying that state, e.g. 01 written above a box signifies that for an order $Q_A Q_B$ then Q_A = '0' and Q_B = '1'. Q_A and Q_B are called the *state variables* since their

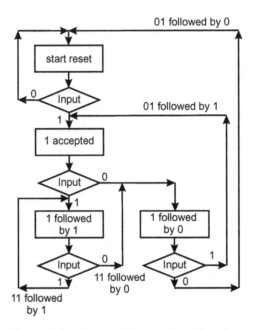

Figure 6.12 ASM chart function for shift register

values describe particular states. For the two-bit shift register the one hot solution is excessively complex, whereas the coded version clearly follows the input data. The bistable state allocation is written beside each box as $Q_A Q_B$, where bistable A takes the input data and bistable B takes the Q_A as its input data (Figure 6.5).

One hot? That's cool!

The ASM chart in Figure 6.13 specifies the control algorithm that we require to implement for a shift register. It specifies how each bistable is set or reset according to the input data and the previous state of the device. Design usually centres on implementation using edge-triggered D-type bistables. For each bistable, the state latched at an active clock transition depends on the data input and the current state, as defined by the bistable outputs. We then need to define the bistable data input D as a function of the input data and the current bistable outputs. This is achieved using a *present state/next state chart*, which tabulates these inputs in a manner similar to

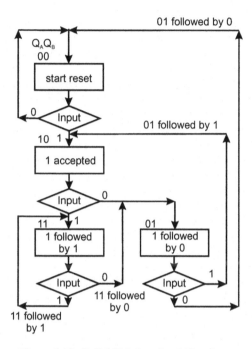

Figure 6.13 Full ASM chart for shift register

a truth table. The transition between present and next state occurs at the clock edge. This is when the next state then becomes the present state of the device.

Often the full set of codes offered by a number of bistables is not needed to implement the full set of states. Should the ASM chart contain only five states, this will require three bistables at a minimum. The bistables offer eight combinations so there will be three combinations, or states, which are not allocated: these are called *unused states*. It is usually better to assign unused states to enter the sequence, so that if they are entered inadvertently (via noise on the inputs, say) then the operation of the device will not be fully corrupted since the sequence will return to that specified in the design. Unused states can be specified to rejoin the sequence at a point that confers a better minimised result. Better minimisation can also be achieved by changing the state allocations, that is, the codes used to represent each state. These points will be illustrated later; we shall first consider the basic implementation.

The present state/next state chart for the two-bit shift register is derived from the ASM chart as:

| Present State | | | Next State | |
| Bistable Outputs | | Input Data | Bistable Inputs | |
Q_A	Q_B	I	D_A	D_B
0	0	0	0	0
0	0	1	1	0
0	1	0	0	0
0	1	1	1	0
1	0	0	0	1
1	0	1	1	1
1	1	0	0	1
1	1	1	1	1

These states cover all those possible for two bistables, so there are no unused states. The method of implementing this is to determine the conditions for which the bistable is set. This is directly equivalent to truth-table extraction of combinational logic implementation. Bistable A is set for $D_A =$ '1'. There are four conditions for this, and these can then be minimised:

$$D_A = \overline{Q_A} \cdot \overline{Q_B} \cdot I + \overline{Q_A} \cdot Q_B \cdot I + Q_A \cdot \overline{Q_B} \cdot I + Q_A \cdot Q_B \cdot I = I$$

Similarly, bistable B can be observed to be set after Q_A has been set, giving the expected extraction

$$D_B = Q_A$$

This gives the expected implementation, identical with that in Section 6.1.4. As with combinational logic implementation, it is often prudent to minimise logic extraction using a K-map. These tenets are best illustrated by application and we shall now design a basic sequential system, a counter.

6.2.2 *Digital counter design: illustrating aspects of ASM design*

6.2.2.1 *Basic counter design: ASM charts in practice*

A counter is a device that has a sequence of states related to count values. A counter can therefore be used to control the sequence of events in a

system; at specified states, chosen events are activated or enabled, and the system is then controlled through the state sequence. Another application of a counter is to provide a delay of a specified number of clock cycles. This can give a very accurate method to provide a delay of a fixed time equal to the number of counted clock events.

Binary counters progress through a binary sequence and the length of this sequence is the *modulus*. This is similar to arithmetic systems where we also talk of modulus arithmetic. A Modulo-8 counter has eight discrete states and is also called a *divide-by-8 counter*. In terms of a binary implementation, a divide-by-8 counter can be implemented using three bistables, since these can be used to encode the eight distinct states. The ASM chart for a divide-by-8 counter comprises eight states and the state machine progresses from state to state at each clock edge. In ASM chart design, the state assignment is arbitrary; any bistable combination can be chosen so long as it is unique for each state. In binary counters the coding is usually chosen to reflect a binary-coded count sequence. Three bistables are used to encode the eight states of a Modulo-8 counter, labelled state 0 to state 7 in Figure 6.14.

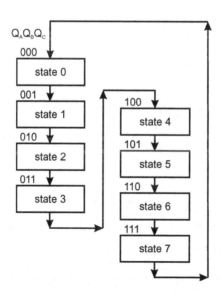

Figure 6.14 ASM chart for Modulo-8 counter

The present state/next state chart derived from the ASM chart in Figure 6.14 is:

Present State Bistable Outputs			Next State Bistable Inputs		
Q_A	Q_B	Q_C	D_A	D_B	D_C
0	0	0	0	0	1
0	0	1	0	1	0
0	1	0	0	1	1
0	1	1	1	0	0
1	0	0	1	0	1
1	0	1	1	1	0
1	1	0	1	1	1
1	1	1	0	0	0

The extraction for implementation is then for D_A and D_B

$$D_A = \overline{Q}_A \cdot Q_B \cdot Q_C + Q_A \cdot \overline{Q}_B \cdot \overline{Q}_C + Q_A \cdot \overline{Q}_B \cdot Q_C + Q_A \cdot Q_B \cdot \overline{Q}_C$$

$$= \overline{Q}_A \cdot Q_B \cdot Q_C + Q_A \cdot \overline{Q}_B + Q_A \cdot \overline{Q}_C$$

$$D_B = \overline{Q}_A \cdot \overline{Q}_B \cdot Q_C + \overline{Q}_A \cdot Q_B \cdot \overline{Q}_C + Q_A \cdot \overline{Q}_B \cdot Q_C + Q_A \cdot Q_B \cdot \overline{Q}_C$$

$$= \overline{Q}_B \cdot Q_C + Q_B \cdot \overline{Q}_C$$

D_C is set only when Q_C is LOW and so

$$D_C = \overline{Q}_C$$

The circuit for a synchronous divide-by-8 counter is shown in Figure 6.15 using an XOR gate for the input D_B, just for simplicity. Sequential circuits can become very complex, though viewing the outputs via a logic simulator is very convincing indeed.

Figure 6.15 Circuit for Modulo-8 counter

The counter's operation is partly depicted in the filmstrip in Figure 6.16. Note that to see the binary count sequence in Figure 6.16, the circuit outputs need to be viewed in reverse order so that at clock 2 the circuit outputs are viewed as $Q_C Q_B Q_A = 100$, whereas the count sequence value is actually $Q_A Q_B Q_C = 001$.

Figure 6.16 Operation of Modulo-8 counter

6.2.2.2 *Gray code counters: ASM implementation minimisation*

The state allocation can also reduce the combinational next state logic. A system commonly used is a *Gray code* where one bit only changes between adjacent states in the sequence. Since one bit only changes between states, it implies that the driving logic to cause a change of state is much reduced. A Gray code for a three-bit counter is:

Code 000 \rightarrow 001 \rightarrow 011 \rightarrow 010 \rightarrow 110 \rightarrow 111 \rightarrow 101 \rightarrow 100
Decimal 0 1 3 2 6 7 5 4

Note that the last one, the eighth code, 100 returns to code 0, 000 so it is a complete Gray code. These give eight distinct states for three bits, but do not reflect a binary-coded system. The most important part of choosing state codes is that each code is unique. This code is used to reduce the amount of logic needed to implement the design. Should a particular state need to be determined, a decoder is required to interpret the bit pattern. There are other Gray codes possible, and this changes the complexity of the result. For a counter, this Gray code can be used in the state allocation, as shown in Figure 6.17.

It is called a Gray code after a patent by Frank Gray in 1953. He was a worker in the early days of television. Apart from digital design, one main use of a Gray code is in shaft encoders where its use avoids errors consequent to sensor misalignment when using a pure binary code. As a reflecting binary code, its first use was in television.

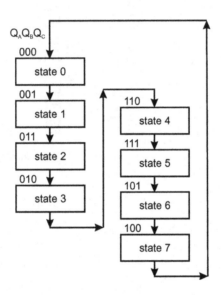

Figure 6.17 ASM chart for Modulo-8 counter using Gray code

The present state/next state chart derived from this ASM chart is:

Present State			Next State		
Q_A	Q_B	Q_C	D_A	D_B	D_C
0	0	0	0	0	1
0	0	1	0	1	1
0	1	1	0	1	0
0	1	0	1	1	0
1	1	0	1	1	1
1	1	1	1	0	1
1	0	1	1	0	0
1	0	0	0	0	0

The extraction for implementation is then for D_A, D_B, D_C is, by minimisation,

$$D_A = Q_B \cdot \overline{Q}_C + Q_A \cdot Q_C; \; D_B = \overline{Q}_A \cdot Q_C + Q_B \cdot \overline{Q}_C;$$
$$D_C = \overline{Q}_A \cdot \overline{Q}_B + Q_A \cdot Q_B$$

This is slightly simpler than the earlier extraction for a binary-coded counter since, even though the expression for D_C is more complicated, the implementation for D_A does not now require any three-input gates. This can be seen in its implementation, which is shown in Figure 6.18. This is the Logisim circuit and the counter is in State 2 where the outputs are in state $Q_A Q_B Q_C = 011$. In this circuit the bistables are positive edge-triggered D-types, though the resolution is not so clear. Fire up Logisim and watch it go!

Figure 6.18 Modulo-8 counter using Gray code (via Logisism)

6.2.2.3 *Further counter design: alternative sequences and unused states*

We can also design counters that, for example, provide either divide-by-3 or divide-by-5 functionality, according to an input control signal. If this signal is '0' for a Modulo-5 counter, and '1' for a Modulo-3, we shall call it $\overline{5}/3$. The ASM chart for this counter then contains five states, of which three are the states in the Modulo-3 counter (when $\overline{5}/3$ is '1'), and two are only entered in the divide-by-5 sequence (when $\overline{5}/3$ is '0') (Figure 6.19).

Three states are not used in the count sequences. In the ASM chart these can be allocated so that all bistables reset at the next clock event, joining

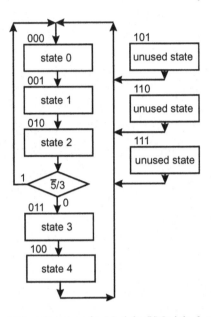

Figure 6.19 ASM chart for Modulo-5/Modulo-3 counter

the sequence at the first count state. Should any of the unused states be entered accidentally, then the device will restart and correct operation will ensue. Unused states can actually be assigned for minimisation, as shown in Section 6.2.2.4.

Present State			Input	Next State			
Q_A	Q_B	Q_C	$\overline{5}/3$	D_A	D_B	D_C	
0	0	0	X	0	0	1	First state of both sequences
0	0	1	X	0	1	0	Second state of both sequences
0	1	0	0	0	1	1	Continue with Modulo-5 sequence
0	1	0	1	0	0	0	Restart Modulo-3 sequence
0	1	1	X	1	0	0	Fourth state of Modulo-5 count
1	0	0	X	0	0	0	Fifth and final state of Modulo-5
1	0	1	X	0	0	0	Reset unused state
1	1	0	X	0	0	0	Reset unused state
1	1	1	X	0	0	0	Reset unused state

> *The unused states could enter the sequence at the point where the control signal is tested, or indeed at any point within the sequence. This may be defined in the specification.*

The extraction is more complicated in this example, so we shall use a K-map to ease the minimisation process.

D_A

$Q_A Q_B$ \ $Q_C \overline{5}/3$	00	01	11	10
00	0	0	0	0
01	0	0	1	1
11	0	0	0	0
10	0	0	0	0

D_B

$Q_A Q_B$ \ $Q_C \overline{5}/3$	00	01	11	10
00	0	0	1	1
01	1	0	0	0
11	0	0	0	0
10	0	0	0	0

D_C

$Q_A Q_B$ \ $Q_C \overline{5}/3$	00	01	11	10
00	1	1	0	0
01	1	0	0	0
11	0	0	0	0
10	0	0	0	0

and so the extraction then reduces to:

$$D_A = \overline{Q}_A \cdot Q_B \cdot Q_C$$

$$D_B = \overline{Q}_A \cdot \overline{Q}_B \cdot Q_C + \overline{Q}_A \cdot Q_B \cdot \overline{Q}_C \cdot \overline{5}/3$$

$$D_C = \overline{Q}_A \cdot \overline{Q}_B \cdot \overline{Q}_C + \overline{Q}_A \cdot \overline{Q}_C \cdot \overline{\overline{5}/3}$$

and the implementation is shown in Figure 6.20.

Figure 6.20 Circuit for Modulo-5/Modulo-3 counter

Clearly, the signal $\overline{5}/3$ could be defined in inverted form, thus saving one inverter.

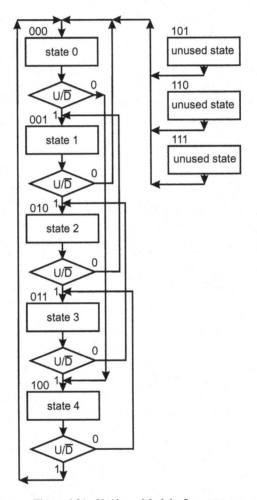

Figure 6.21 Up/down Modulo-5 counter

6.2.2.4 *Minimising sequential logic*

As ever, there is a natural desire to *minimise sequential circuits*, to save money and improve performance. We have already considered many of

the techniques that achieve this: we can change the state sequence for minimisation; we have assigned unused states; and we can assign these for minimisation purposes. In software design systems this is achieved by programming and an exhaustive search of all possibilities to ensure that the minimum implementation is achieved. Here, we need some by-hand tools: step up the K-map!

As a vehicle for illustration, we shall consider an up/down Modulo-5 counter. The control signal is named U/\overline{D} and when HIGH the counter counts **up**, Modulo-5, and when U/\overline{D} is LOW the counter counts **down**, again Modulo-5. This is described by the ASM chart in Figure 6.21, on which the state coding is not shown as we shall be using that next. Our first approach will be to use a K-map for minimisation. We shall first assign a binary code to the states:

State	Code	State	Code	State	Code	State	Code
0	000	2	010	4	100	6 (unused)	110
1	001	3	011	5 (unused)	101	7 (unused)	111

Our present state/next state chart is then, for a D-type implementation:

Present State			Input	Next State		
Q_3	Q_2	Q_1	U/\overline{D}	D_3	D_2	D_1
0	0	0	0	1	0	0
0	0	0	1	0	0	1
0	0	1	0	0	0	0
0	0	1	1	0	1	0
0	1	0	0	0	0	1
0	1	0	1	0	1	1
0	1	1	0	0	1	0
0	1	1	1	1	0	0
1	0	0	0	0	1	1
1	0	0	1	0	0	0
1	0	1	X	0	0	0
1	1	0	X	0	0	0
1	1	1	X	0	0	0

Digital Electronics: A Primer

By drawing K-maps for each D input:

D_3

$Q_1 U/\overline{D}$ \ Q_3Q_2	00	01	11	10
00	1	0	0	0
01	0	0	0	0
11	0	1	0	0
10	0	0	0	0

D_2

$Q_1 U/\overline{D}$ \ Q_3Q_2	00	01	11	10
00	0	0	0	1
01	0	1	0	0
11	1	0	0	0
10	0	1	0	0

D_1

$Q_1 U/\overline{D}$ \ Q_3Q_2	00	01	11	10
00	0	1	0	1
01	1	1	0	0
11	0	0	0	0
10	0	0	0	0

This is not too good for minimisation since we can draw few loops greater than a single cell. The extraction obtained gives the equations implemented in the circuit of Figure 6.22.

$$D_3 = \overline{Q}_3 \cdot \overline{Q}_2 \cdot \overline{Q}_1 \cdot \overline{U/\overline{D}} + \overline{Q}_3 \cdot Q_2 \cdot Q_1 \cdot U/\overline{D}$$

$$D_2 = \overline{Q}_3 \cdot \overline{Q}_2 \cdot Q_1 \cdot U/\overline{D} + \overline{Q}_3 \cdot Q_2 \cdot \overline{Q}_1 \cdot U/\overline{D}$$

$$+ \overline{Q}_3 \cdot Q_2 \cdot Q_1 \cdot \overline{U/\overline{D}} + Q_3 \cdot \overline{Q}_2 \cdot \overline{Q}_1 \cdot \overline{U/\overline{D}}$$

$$D_1 = \overline{Q}_3 \cdot Q_2 \cdot \overline{Q}_1 + \overline{Q}_3 \cdot \overline{Q}_1 \cdot U/\overline{D} + Q_3 \cdot \overline{Q}_2 \cdot \overline{Q}_1 \cdot \overline{U/\overline{D}}$$

The circuit for this is given in Figure 6.22. This has been taken from Logisim and labels added (so the labels are slightly different from usual). This is so there are no drawing errors in this complicated circuit. Roll on software descriptions! (Go ahead and check it out in Logisim too — and you can see how it works.)

If we were to use a Gray code state allocation, minimisation would certainly improve. We will always have a situation where two bits change at the same time, since the modulus of the counter is uneven. If it was an even modulus, e.g. Modulo-6, we would have a proper Gray code, but

Figure 6.22 Modulo-5 up/down binary counter

that is not the case here. With one Gray code, the present state/next state becomes:

Q_3	Q_2	Q_1	U/\overline{D}	D_3	D_2	D_1
\multicolumn{3}{c}{Present State}	Input	\multicolumn{3}{c}{Next State}				

Present State			Input	Next State		
Q_3	Q_2	Q_1	U/\overline{D}	D_3	D_2	D_1
0	0	0	0	1	0	1
0	0	0	1	0	0	1
0	0	1	0	0	0	0
0	0	1	1	0	1	1
0	1	1	0	0	0	1
0	1	1	1	1	1	1
1	1	1	0	0	1	1
1	1	1	1	1	0	1
1	0	1	0	1	1	1
1	0	1	1	0	0	0
0	1	0	X			
1	0	0	X			
1	1	0	X			

And, leaving the unused states as unspecified (we shall get to those in a moment) the K-maps are then:

D_3

Q_3Q_2 Q_1U/\overline{D}	00	01	11	10
00	1			
01	0			
11	0	1	1	0
10	0	0	0	1

D_2

Q_3Q_2 Q_1U/\overline{D}	00	01	11	10
00	0			
01	0			
11	1	1	0	0
10	0	0	1	1

D_1

Q_3Q_2 Q_1U/\overline{D}	00	01	11	10
00	1			
01	1			
11	1	1	1	0
10	0	1	1	1

By the extraction, D_2 is much reduced, but D_3 and D_1 are still complex. Ignoring the unused states (or treating them all as 0) we obtain:

$$D_3 = \overline{Q}_3 \cdot \overline{Q}_2 \cdot \overline{Q}_1 \cdot \overline{U/\overline{D}} + Q_2 \cdot Q_1 \cdot U/\overline{D} + Q_3 \cdot \overline{Q}_2 \cdot Q_1 \cdot U/\overline{D}$$

$$D_2 = \overline{Q}_3 \cdot Q_1 \cdot U/\overline{D} + Q_3 \cdot Q_1 \cdot \overline{U/\overline{D}}$$

$$D_1 = \overline{Q}_3 \cdot \overline{Q}_2 \cdot \overline{Q}_1 + \overline{Q}_3 \cdot Q_1 \cdot U/\overline{D} + Q_2 \cdot Q_1 + Q_3 \cdot Q_1 \cdot \overline{U/\overline{D}}$$

Now we can allocate the unused states. Note that these are not don't-care states, since we definitely **do care**. We don't want a situation where one unused state goes to another, and that one returns to the first unused state. If that were the case, on entering an unused state, the system would cycle between unused states. We want the unused states to rejoin the sequence. If we have a free choice as to where they return, we can minimise the implementation further. By inspection of the K-maps, we can choose the unused state allocation so as to make the loops larger, hence minimising further. We then have:

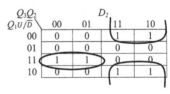

We need to check that the unused states rejoin the sequence.

Present (Unused) State			Input	Next State		
Q_3	Q_2	Q_1	U/\overline{D}	D_3	D_2	D_1
0	1	0	0	1	0	1
0	1	0	1	1	0	1
1	0	0	0	0	1	1
1	0	0	1	0	0	1
1	1	0	0	1	1	1
1	1	0	1	1	0	1

This is fine, since no unused state goes to another unused state; all unused states rejoin the sequence. Our extraction is then:

$$D_3 = \overline{Q}_1 \cdot \overline{U/\overline{D}} + Q_2 \cdot U/\overline{D} + Q_3 \cdot \overline{Q}_2 \cdot U/\overline{D}$$

$$D_2 = \overline{Q}_3 \cdot Q_1 \cdot U/\overline{D} + Q_3 \cdot \overline{U/\overline{D}}$$

$$D_1 = \overline{Q}_1 + Q_2 + Q_3 \cdot \overline{U/\overline{D}} + \overline{Q}_3 \cdot U/\overline{D}$$

Figure 6.23 Minimised Modulo-5 up/down counter

This is clearly much smaller, especially by comparison with the circuit in Figure 6.22. It is the same basis as Figure 6.1 and the reduced complexity is emphasised by the lack of driving logic. Note that it's the same speed as the AND/OR structures are identical. It is just a lot cheaper in terms of logic, and it will be more reliable, too, since there is less to go wrong.

By changing the state allocation and by choosing the unused states carefully, we can minimise the implementation. But this is not necessarily the smallest circuit. There are other Modulo-5 Gray codes, and there are other unused state allocations. To check these out would be a very lengthy procedure. That's what computers are good at. In fact, they can minimise automatically and exclusively search all the combinations for us, just in the flash of a multicore N GHz system. That means that we have to program the system, and that is just the way it is done these days.

6.2.2.5 *Implementing counters using* T-*type bistables*

We can also use T-type bistables for implementation. The procedure uses the same ASM chart and changes the present state/next state chart so as to derive conditions for which the bistables toggle (as opposed to D-type implementation where we derive conditions for which the bistables are set). We then derive equations for implementation using T-types. Eventually this reduces to the same equations for a D-type implementation, as we shall show.

We shall illustrate this for a Modulo-8 counter from Section 6.2.2.1 with the ASM Chart shown in Figure 6.14. First, we change the present state/next state chart for T-type implementation. This requires that we determine the conditions for which the bistable will toggle. So we have shown the present state/next state transitions and in the last three columns the T signals are '1' when the Q outputs change from '0' to '1'. We have noted the present state as Q and the next state as Q^+.

Present State Bistable Outputs			Next State Bistable Outputs			Toggle		
Q_A	Q_B	Q_C	Q_A^+	Q_B^+	Q_C^+	T_A	T_B	T_C
0	0	0	0	0	1	0	0	1
0	0	1	0	1	0	0	1	1
0	1	0	0	1	1	0	0	1
0	1	1	1	0	0	1	1	1
1	0	0	1	0	1	0	0	1
1	0	1	1	1	0	0	1	1
1	1	0	1	1	1	0	0	1
1	1	1	0	0	0	1	1	1

Here, the toggle signals T inform us when the outputs change: '1' for a change and '0' for a latch. We can then use combinational logical analysis to determine which collections of the current outputs Q give us the T inputs (Figure 6.24). The equations are then:

$$T_A = Q_B \cdot Q_C$$
$$T_B = Q_C$$
$$T_C = 1$$

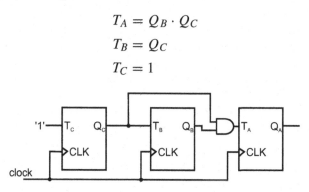

Figure 6.24 Modulo-8 counter via T-type bistables

These are, of course, the same as the D-type implementation. Using the equation for a T-type bistable from Section 5.5.3,

$$D = \bar{T} Q + T \bar{Q}$$

For the D-type implementation using these equations (and De Morgan's law)

$$D_A = \overline{(Q_B \cdot Q_C)} Q_A + (Q_B \cdot Q_C) \bar{Q}_A$$
$$= \bar{Q}_A \cdot Q_B \cdot Q_C + Q_A \cdot \bar{Q}_B + Q_A \cdot \bar{Q}_C$$
$$D_B = \bar{Q}_C Q_B + Q_C \bar{Q}_B$$
$$D_C = 0 \cdot Q_C + 1 \cdot \bar{Q}_C = \bar{Q}_C$$

This is then the same as the D-type implementation in Section 6.2.2.1, as expected. It just provides a different way to achieve the design. The reason some people favour using T-type implementations is because the design is straightforward (and then you whack the design into a minimiser). For an N-bit counter, at each stage we derive the T input by ANDing all the Q outputs from the preceding bistables, which, in general, for a set of N bistables with the least significant bit Q_0 is

$$T_n = Q_{n-1} \cdot Q_{n-2} \cdot \ldots \cdot 1 \quad \forall n \in 0, 1..N - 1$$

This only has currency when *JK*-type bistables are available, since the design just uses a sequence of larger AND gates (since a *T*-type bistable can be made from a *JK*-type simply by connecting the *J* and the *K* inputs together, as in Section 5.5.3). These days, the *D*-type bistable is ubiquitous, so this approach is of much less interest now, though it does give a simple design route. In fact, you were earlier promised that it is possible make a *JK*-type from a *D*-type. So let's do it now.

6.2.2.6 *Converting a* D-*type into a* JK-*type*

First step: an ASM chart (Figure 6.25). There are two inputs, *J* and *K*, and a bistable with two states: set and reset (and the bistable can latch or toggle these states). For space, we shall insert both inputs within the decision box. The notation *J*, $K = 0, 0 + 0, 1$ means that branch is selected when *J*,*K* are both zero or are $J = 0$ and $K = 1$.

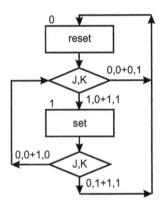

Figure 6.25 ASM chart for a *JK*-type

The present state/next state is then:

Inputs		Present State	Next State
J	*K*	*Q*	*D*
0	0	0	0
0	0	1	1
0	1	X	0
1	0	X	1
1	1	0	1
1	1	1	0

From this, we obtain (via a K-map if you prefer)

$$D = J \cdot \bar{Q} + \bar{K} \cdot Q$$

So we can now make a *JK* from a *D*-type. Note that the earlier section concerned how to make a *D*-type into a *T*-type. It is different here since we seek to give a *D*-type the full *JK*-type functionality, which has set and reset as well as latch and toggle. The complexity is about the same: just a couple of gates are needed to give the *D*-type the required functionality.

6.2.3 *Translating ASM charts into a software specification*

In order to translate an ASM chart into software we must translate the present state/next state chart into software. That is how the whole process started earlier. Essentially we need to specify that position in the state sequence, and any tested inputs define the next state in the sequence. The present state/next state chart comes from the ASM chart as does the software specification. The present state/next state chart includes all possible states and values of the inputs. These could be tested using multiple IF ... THEN ... ELSE statements but to reduce complexity we will use a CASE statement. The CASE construct requires variables to be tested and these will be the current outputs of the bistables expressed in vector form. We then enter a new state, equivalent to assigning new values to each bistable according to any inputs tested in the current state. This assumes that the implementation is specifically targeted at *D*-type bistables. The assignment operator is what happens at a clock edge, namely that a new state is entered.

The present state/next state chart for a Modulo-8 counter and its software specification are

Q_A	Q_B	Q_C	D_A	D_B	D_C	CASE (Q[2..0])
\multicolumn						

Present State			Next State			CASE (Q[2..0])
Q_A	Q_B	Q_C	D_A	D_B	D_C	BEGIN
0	0	0	0	0	1	0 : Q[2..0]=1
0	0	1	0	1	0	1 : Q[2..0]=2
0	1	0	0	1	1	2 : Q[2..0]=3
0	1	1	1	0	0	3 : Q[2..0]=4
1	0	0	1	0	1	4 : Q[2..0]=5
1	0	1	1	1	0	5 : Q[2..0]=6
1	1	0	1	1	1	6 : Q[2..0]=7
1	1	1	0	0	1	7 : Q[2..0]=0
						END

and the software specification for the $\overline{5}/3$ counter derived from its present state/next state chart includes testing the $\overline{5}/3$ signal to determine which sequence the counter should enter and is

Q_A	Q_B	Q_C	$\overline{5}/3$	D_A	D_B	D_C
0	0	0	X	0	0	1
0	0	1	X	0	1	0
0	1	0	0	0	1	1
0	1	0	1	0	0	0
0	1	1	X	1	0	0
1	0	0	X	0	0	0
1	0	1	X	0	0	0
1	1	X	X	0	0	0

With header rows: Present State columns (Q_A Q_B Q_C $\overline{5}/3$), Next State columns (D_A D_B D_C).

```
CASE (Q[2..0])
  BEGIN
    0:  Q[2..0]=1
    1:  Q[2..0]=2
    2:  IF (5_3) THEN Q[2..0]=0
                 ELSE Q[2..0]=3
    3:  Q[2..0]=4
    4:  Q[2..0]=5
    6,7,8: Q[2..0]=0
  END
```

This, of course, presumes that we know the precise state allocation. If this is to be chosen by the system (for minimisation purposes) then our state machine description will be of the form:

```
CASE state
  BEGIN
  first: IF up THEN state:=second ELSE state:=last
  second: IF up THEN state:=third ELSE state:=first
  third: IF up THEN state:=fourth ELSE state:=second
  fourth: IF up THEN state:=last ELSE state:=third
  last: IF up THEN state:=first ELSE state:=fourth
  END
```

It should be checked that the unused states have been allocated correctly (in some systems unspecified states are don't-care by default). Clearly, the ASM chart is not needed since the system is fully defined in software. This has the advantage that the system can be verified: it can be checked whether the operation completely fulfils the circuit description. By this argument the state machine specification is done completely in software and we don't need anything else. That is how these circuits are implemented now, and the minimisation and choice of state values can be optimised by a computer in milliseconds, given its awesome processing power.

Don't be tempted to think that since there is software and systems already on the market that can do all this design for you, then you don't need to learn it. Someone has to develop the system. That could be you. Learn!

6.3 Further Concepts in Sequential Design

6.3.1 *Further ASM chart design*

Traditionally, an output generated in a particular state is written within the state symbol and the output will exist for the duration of that state. Outputs are often required which exist for a certain state, given that a certain condition is met. These are termed *conditional outputs*, since they depend on (are conditional on) the inputs, and are symbolised as shown in Figure 6.26.

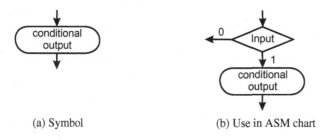

(a) Symbol (b) Use in ASM chart

Figure 6.26 Conditional outputs in ASM charts

Conditional outputs naturally occur in ASM charts after an input, or condition, has been tested. Given that we are in one state, A, then the conditional output will exist only if we are in state A and the input tested in state A is FALSE. The outputs that exist in a particular state are actually termed *unconditional outputs*.

6.3.2 *Other sequential specifications: Moore and Mealy charts*

There are alternative state machine descriptions and these include the *Moore* and *Mealy* machines. These employ *state transition diagrams*, which, like ASM charts, describe sequence of states in a sequential system. The traditional Moore machine (Figure 6.27[a]) is a state diagram

containing only unconditional outputs and no conditional ones. On the other hand, the traditional Mealy machine has conditional outputs and a Mealy state machine can therefore have more than one output value in a state (Figure 6.27[b]).

The states are symbolised using circles and the transitions are given by the links between the states. The state number and the output associated with that state are written within the circle denoting the state. The lines connecting states indicate the input conditions for a change between states and the status of a particular input is written beside the line.

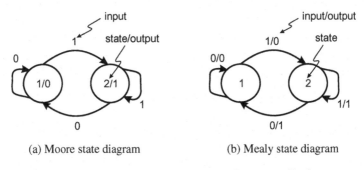

(a) Moore state diagram (b) Mealy state diagram

Figure 6.27 Alternative state diagrams for *D*-type flip-flop

In a Mealy state diagram (with conditional outputs) the lines connecting states have not only the inputs, but also the outputs. These diagrams can become rather complex for larger systems. Since the ASM method is a state machine employing both types of output, then the ASM technique includes both cases and has therefore found increasing popularity for design purposes.

6.3.3 *A modern view of sequential design*

So far, we have rather stuck with convention: for sequential design, first draw a transition diagram and extract the next state equations from the present state/next state chart. Given that modern implementation is by software, the question is then whether it is all necessary. The one part of dubious value is the one where there is a selection of techniques: do we actually need a state transition diagram at all? As we have seen, there are Moore and Mealy charts that are combined in the ASM method. Each

method leads to the same present state/next state chart (it must do) so that rather underlines the point that the result implies that there is no favoured choice of method.

We should also consider implementation. Modern systems are so complex that the design is software based. This allows for the use of optimisation and for minimisation. It also allows for validation: we can prove that the design fully meets the specification. In later studies you will find test and testability where you can design for and check the system's operation. For simple designs, these can be achieved by hand, but for complex designs this needs software. As such, everything rests on the present state/next state chart, since this is the complete description of the circuit's operation.

These two factors lead to the conclusion that the choice of system specification is personal rather than optimal. A purely software specification is a way forward since the design can be validated and designed for testability. It can easily be simulated too, so the system's correct operation can be assured. This is just one view. Specification is always needed and some prefer a more traditional approach. In a way it is a bit like the Karnaugh map: nice for basic design and simple systems but completely impractical for large numbers of inputs or states. As for static hazards, the advice earlier was to expect timing as part of life and design around it. In the practical design of sequential logic systems, when faced with a real system with hundreds of inputs and states, designers will go straight to some form of software specification for that is — rather bluntly — the only way that is practicable.

6.4 Asynchronous Sequential Logic

6.4.1 *Basic asynchronous counters*

We have so far considered *synchronous sequential design* only. In such systems all timing is specifically linked to the clock and the same clock event is used by all bistables. As a simpler explanation, everything changes at the same time and it can be inferred that this leads to greater reliability. There are also *asynchronous sequential systems* where timing is not specifically linked to the clock. Asynchronous systems move through a sequence of states but outputs do not all change at the same time. The main advantages associated with asynchronous circuitry are a reduction in logic (and hence

cost), which can confer an increase in speed. These advantages are only accrued with potential for unreliability and, since few engineers aim to build systems with inherent faults, it follows that asynchronous systems are usually only used in very special applications where their advantages are so important that a designer is willing to tolerate (and compensate for) any potential faults.

There are *asynchronous state machines* as well as synchronous ones. These are not synchronised to a clock but driven by changes in the inputs that cause changes between states. The design methods are more complex than for their synchronous counterparts and the results are more complex to analyse, but can achieve greater speed and minimisation in implementation.

The contrast between asynchronous and synchronous systems is perhaps best illustrated by counter design: asynchronous counters are simple but can be prone to error; synchronous counters are more complex, harder to design, but are more reliable. The asynchronous counters presented here are not asynchronous state machines, but are asynchronous systems that have bistable triggering signals, which come from different sources.

An important element in asynchronous counter design is the toggle-type or T-type bistable (Figure 6.28). In a T-type, covered earlier in Section 5.5.3, according to the status of a control signal, the outputs either toggle (as in the toggle state of a JK bistable), or latch.

T	CLK	Q_{n+1}	\overline{Q}_{n+1}
X	0	Q_n	\overline{Q}_n
X	1	Q_n	\overline{Q}_n
0	↑	Q_n	\overline{Q}_n
1	↑	\overline{Q}_n	Q_n

Figure 6.28 T-type bistable revisited

This can be easily constructed using a D-type, since $D = T\overline{Q} + \overline{T}Q$. The T-type outputs change state on active clock transitions when the T input is HIGH, so an input clock frequency is then divided by two to give a divide by two or Modulo-2 counter (it is counting 0,1,0,1,0,1, ...). For a positive edge-triggered T-type this is as shown in Figure 6.29.

Figure 6.29 Toggling a *T*-type; frequency division by two

If we connect more *T*-type bistables in series, connecting the clock for a following stage to the output of the previous stage, then at each stage we increase frequency division of the input clock frequency by two. Three bistables connected together as in Figure 6.30 give frequency division by eight. Note that here we are using **negative** edge triggering. If the active clock event were positive, the system would count differently.

Figure 6.30 Asynchronous divide-by-8 (Modulo-8) counter

The timing diagram (Figure 6.31) shows this frequency division in practice. If we **visualise** the output states as a binary number $Q_3 Q_2 Q_1$, we can see the counter cycling between 0 to 7 and back again.

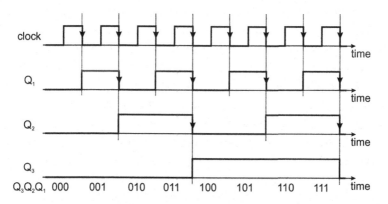

Figure 6.31 Timing for divide-by-8 (Modulo-8) counter

The system is, however, **asynchronous**, since not all clock signals are connected to the same clock input. Since the Q output of one stage provides the clock for the following stage, the second stage can only change two propagation delays after the clock, one propagation delay after the first output changes. This provides a clock that ripples through the system and for this reason this counter is often called a *ripple counter*. Because of the ripple effect, the real timing diagram is then quite different from its idealised version.

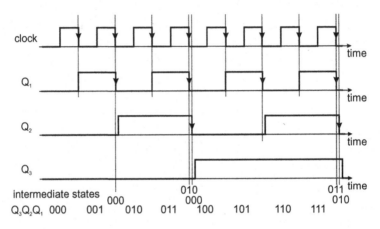

Figure 6.32 Real timing for divide-by-8 counter with erroneous states

The decoded outputs are then $Q_3Q_2Q_1$ and intended state outputs and some intermediate states are given in Figure 6.32. The intermediate states are inherent in this design since each bistable output provides the clock signal for the following stage, so the output of the bistable must fall before the next bistable changes its value (the bistable output changes when the propagation time has elapsed). Since the outputs of counters are often decoded (for example, in sequence control), if one device is enabled for a count of 0 ($Q_3 = Q_2 = Q_1 = $ '0') and another device enabled at the fifth cycle ($Q_3 = $ '1', $Q_2 = Q_1 = $ '0'), then the first device will be enabled for a full clock period at first and then later for a very short time when the device changes between $Q_3 = $ '0', $Q_2 = Q_1 = $ '1' to $Q_3 = $ '1', $Q_2 = Q_1 = $ '0'. In fact, the count sequence of eight is actually a count sequence of 14, though six of these are unwanted erroneous (intermediate) states of very short duration. This is usually completely unwanted and incorrect system

operation would doubtless ensue. A device enabled $Q_3 = Q_2 = Q_1 =$ '0' will also be erroneously activated for a short time between the second and the third, the fourth and the fifth, and between the sixth and seventh clock cycles. This is clearly unacceptable in many systems.

The main advantage of these asynchronous counters is **simplicity**. The asynchronous counter uses just T-type bistables with no further combinational logic. This is much simpler than its synchronous equivalent. The price paid for this simplicity is in performance. A fully synchronous counter is usually preferred in sequential systems since it has no unwanted intermediate states and all outputs change at the same time. There is one common application of a ripple counter. In digital watches the original clock frequency is provided by a quartz crystal because the frequency can be controlled very precisely. Since the quartz crystal is physically very small the clock frequency is very high (32.768 kHz) and needs division by a factor of 2^{15} to achieve a 1 Hz (or faster) clock signal. This is often implemented using a ripple counter because of its simplicity (and hence low cost) and because the outputs do not need to be decoded to sequence control within the watch.

6.4.2 *Positive edge-triggered asynchronous counter*

Let us now see what happens if we use positive edge-triggered bistables, instead of negative edge-triggered ones. The circuit now becomes as per Figure 6.33.

Figure 6.33 Asynchronous divide-by-8 (Modulo-8) down counter

Let us assume we again start at $Q_1 Q_2 Q_3 = 000$ (all bistables are reset). At the first rising clock edge, Q_1 will toggle and so Q_1 will change from 0 to 1. That means the second bistable receives a rising clock edge and thus it will toggle — Q_2 will change from 0 to 1 (a short time after Q_1 changes from 0 to 1). This then gives a triggering clock input to third bistable, so it will toggle and Q_3 will change from 0 to 1 (again, a short time after Q_2

changes from 0 to 1). That means at the first rising clock edge the system will change from $Q_1 Q_2 Q_3 = 000$ to $Q_1 Q_2 Q_3 = 111$. At the second rising clock edge Q_1 will toggle and so Q_1 will change from 1 to 0. The clock input to the second bistable then receives a falling clock edge and this causes no change so Q_2 remains at 1. Since Q_2 does not change, there is no rising clock edge fed to the third bistable and so Q_3 remains at 1. That means at the second clock edge the system will change from $Q_1 Q_2 Q_3 = 111$ to $Q_1 Q_2 Q_3 = 011$. A similar analysis shows that at the third clock edge the system will change from $Q_1 Q_2 Q_3 = 011$ to $Q_1 Q_2 Q_3 = 101$. Decoding the bistable outputs as $Q_3 Q_2 Q_1$ we find that the sequence is:

Q_3	Q_2	Q_1
0	0	0
1	1	1
1	1	0
1	0	1
1	0	0
0	1	1
0	1	0
0	0	1
0	0	0
1	1	1
:	:	:

This is a down counter. We start at 0 and then go through the states 7, 6, 5, 4, 3, 2, 1, 0 where the sequence restarts and repeats indefinitely.

6.4.3 *Interrupting the count sequence: asynchronous counters Modulus $\neq 2^N$*

Inclusion of the asynchronous inputs to control the state of the bistable naturally leads to an asynchronous circuit. The asynchronicity then arises from the use of the inputs, rather than the clock. The ripple counter derived so far is only good to provide frequency division by 2^N for a system with N bistables. If we want a longer count sequence, we can extend it by a factor of two by including a further bistable element. Earlier we designed a

Modulo-3 and a Modulo-5 synchronous counter. These can be achieved in an asynchronous system by simply terminating the count, by resetting all bistables, when the counter reaches a specified *terminal count*.

Q_2	Q_1
0	0
0	1
1	0
1	1
0	0

 (a) count sequence (b) circuit

Figure 6.34 Asynchronous Modulo-3 counter

If we require a Modulo-3 counter then we want the system to reset when it reaches a terminal count of $Q_1 = Q_2 = $ '1'. We want to reset both bistables when this state is reached, so we need a clear signal derived from $Q_1 = Q_2 = $ '1'. Clear signals are usually active-low so if we NAND together Q_1 and Q_2 we obtain a signal which is LOW when both bistable outputs are HIGH, as shown in Figure 6.34. We can use this to force the count sequence to enter the state $Q_1 = Q_2 = $ '1' and then reset immediately to $Q_1 = Q_2 = $ '0', giving the required Modulo-3 counter.

This circuit will usually work, but only just. This is because the first bistable to reset will cause the clear signal to return to HIGH, its inactive state. If there is a large difference in the *reset propagation times* (the time taken for the clear signal to take effect), then one bistable might not be reset. Usually, the clear signal is very short lived, but sufficient to clear both bistables. Since operation is not guaranteed, it is therefore best to use a fully synchronous counter because it offers guaranteed performance with little extra combinational logic cost. The simplicity would be better demonstrated with a counter with a longer sequence. A synchronous counter Modulo-3 counter can be designed which uses the same logic as the circuit in Figure 6.34, so it is hard to see why anyone would use the circuit of Figure 6.34 (though this is not true for counters with a longer

sequence). It does illustrate that the asynchronous inputs can be used to control operation, but at potential cost of reliability. A, say, Modulo-14 asynchronous counter would have two more bistables only whereas a synchronous Modulo-14 counter would have much more logic as well as two more bistables. It's then a compromise between simplicity/cost and efficacy/practicality.

6.5 Concluding Comments and Further Reading

Design also usually centres on synchronous sequential systems, since these offer greater potential for reliable operation than their asynchronous counterparts. Algorithmic state machine chart design is a major topic in synchronous sequential systems. This is a formal technique eminently suited to the description and implementation of synchronous sequential systems. As a design technique it is less suited to software implementation. Software implementation can provide a clear description of a synchronous sequential system, but with limited facility for the inclusion of asynchronous systems. Using software to specify systems now plays an important part in digital system design. The importance of software specification has increased concurrently with the range of devices which have been specifically developed for software specification: this is *programmable logic*, where software specifies a chip's function and the way that the logic circuits are connected together to achieve the software specification.

The classic text on ASM is Clare (1973), but it is hard to find a copy now. All logic texts will choose one design technique: for more treatment of Moore and Mealy, try Katz and Borriello (2004). Some books actually omit formal design and just aim for a working system, such as Tocci *et al.* (2004), though this is perhaps rather too influenced by legacy technology. The major implementation now is the use of programmable logic and modern texts reflect this, such as Zwolinski (2004) (a Verilog version is Zwolinski [2009]) and Floyd (2002). A combined text is Vahid (2010) (note that the last bit of the title is a bit misleading, and it's a whacking US$133). The design principles will not change with implementation technology. There will be a greater influence of testability and formal specification/verification, and this is consistent with the increased complexity of current architectures. This is the focus of newer design tools.

6.6 Questions

Solutions in Appendix 1.

Q6.1 Design a synchronous Modulo-7 counter and implement it using positive edge-triggered D-type bistables and standard logic gates (AND, OR, or NOT) only.

Q6.2 Design a synchronous Modulo-7 counter for implementation using positive edge-triggered T-type bistables and standard logic gates (AND, OR, or NOT) only.

Q6.3 Design an asynchronous Modulo-7 counter and implement it using positive edge-triggered D-type bistables and standard logic gates (AND, NAND, OR, NOR, or NOT) only.

Q6.4 Compare and contrast the implementation of a synchronous Modulo-7 counter using positive edge-triggered D-types and T-types and an asynchronous Modulo-7 counter using negative edge-triggered D-types.

Q6.5 Express the ASM chart in Figure 6.35 in software using a CASE statement.

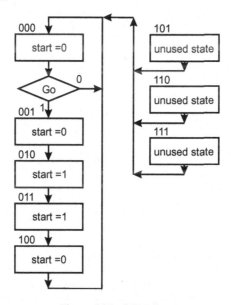

Figure 6.35 ASM chart

Q6.6 Express the ASM chart in the previous question, again in software, but using IF ... THEN constructs only.

Q6.7 The following two circuits are Modulo-7 counters aiming to provide
two (HIGH) pulses of length 0.1 μs spaced by 0.2 μs and 0.3 μs,
given a clock frequency of 10 MHz. The bistables are all positive
edge-triggered D-types and the lower input in Figure 6.36(b) is
the \overline{CLR} input and the D inputs and the \overline{Q} outputs are unlabelled
(the circuits are labelled versions of the Logisim outputs, so you
can also see them operate). The basis of Figure 6.36(b) is the
asynchronous counter in Figure 6.34. Which of the circuits would
you choose for a robust implementation, and why?

(a) Synchronous counter

(b) A synchronous counter

Figure 6.36 Two counters

Q6.8 A system has four input channels — A, B, C, and D — and one output channel. The output channel is selected from one of the input channels in the following sequence:

$$A,B,C,D,C,B,A,B,C,D,C.\ldots \quad \text{et seq.}$$

Two control signals are provided: one to reset the system at any time to start with an output from channel A; the other is a short pulse synchronised to change the output selection. Design a circuit to implement this system and provide a circuit diagram for your design.

Q6.9 Design a system that operates as either a two-bit shift register or as a two-bit binary counter. The signal S/\overline{C} is '1' to select action as a shift register and '0' to select the counter action. Specify your design using an ASM chart and implement it using positive edge-triggered D-type bistables and standard logic gates (AND, OR, or NOT). Given that for the bistables $T_{SU} = 10\,\text{ns}$, $T_H = 2\,\text{ns}$, and $T_{PHL} = 13\,\text{ns}$, and $T_{PLH} = 12\,\text{ns}$, for any combinational gate $T_{PHL} = 11\,\text{ns}$ and $T_{PLH} = 12\,\text{ns}$, estimate the maximum clock frequency at which your circuit will operate reliably.

Q6.10 Design a three-bit bidirectional shift register that shifts its data left if a mode signal M is '0' and right if the mode signal is '1'. The shift register should accept data and shift on every third pulse of a clock signal. Specify the complete circuit using an ASM chart and implement it using positive edge-triggered D-type bistables and standard logic gates (AND, OR, or NOT) only.

Q6.11 Analyse the circuit in Figure 6.37. What signal will appear at the output of the AND gate? Why is the circuit inefficient?

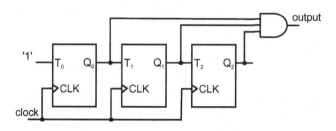

Figure 6.37 Circuit made from T-type bistables

Chapter 7

Number Systems, Coding, and Arithmetic

7.1 Numbers and Coding Systems

7.1.1 *Common number coding systems*

The basic form of a number is a collection of digits arranged into *integer* and *fractional* parts. The digits are arranged in order of significance and numbers are *left justified* when the most significant digit is to the left. A number's arrangement is

$$N = D_n \times r^n + D_{n-1} \times r^{n-1} + \cdots D_1 \times r^1$$
$$+ D_0.D_{-1} \times r^{-1} + \cdots + D_{-m} \times r^{-m}$$

The part of the number to the left of the (decimal) point is an $n + 1$ digit integer and the part to the right is an m digit fraction. The base of the number is called its *radix* and this is denoted by r in the number. Each digit is multiplied by the radix raised to the power associated with that digit. The least significant digit in the integer part is actually $D_0 \times r^0$ and the radix part is omitted, since $r^0 = 1$. In order to avoid confusion, the radix is often used as a subscript to denote the number system used, so 11_{10} denotes 11 in the decimal system, whereas 11_2 denotes a binary number (3_{10}). A *decimal number* is base 10 and so the radix r is 10. The decimal system uses ten symbols, 0, 1, 2...9. The number 1874.28_{10} is then

$$D_3 \times r^3 + D_2 \times r^2 + D_1 \times r^1 + D_0. D_{-1} \times r^{-1} + D_{-2} \times r^{-2}$$
$$= 1 \times 10^3 + 8 \times 10^2 + 7 \times 10^1 + 4. 2 \times 10^{-1} + 8 \times 10^{-2}$$

$$= 1 \times 1000 + 8 \times 100 + 7 \times 10 + 4.\ 2 \times 0.1 + 8 \times 0.001$$

$$= 1874.\ 28_{10}$$

The *binary system* has only two values. The radix is two and the two symbols are 0 and 1. Each digit in a binary number is known as a *binary digit* or *bit*. In *natural binary*, the number 1101.01_2 represents the number

$$D_3 \times 2^3 + D_2 \times 2^2 + D_1 \times 2^1 + D_0.\ D_{-1} \times 2^{-1} + D_{-2} \times 2^{-2}$$

$$= 1 \times 2^3 + 1 \times 2^2 + 0 \times 2^1 + 1.\ 0 \times 2^{-1} + 1 \times 2^{-2}$$

If you want to be mega pedantic, it's now called a 'binary' point, not a decimal point. When I (once) called it a binary point in lectures, some guy at the front muttered 'get a life'. Good point (sic).

And in decimal

$$1101.01_2 = 1 \times 8 + 1 \times 4 + 0 \times 2 + 1.0 \times 0.5 + 1 \times 0.25 = 13.25_{10}$$

Binary coded decimal (BCD) uses decimal digits coded using the binary system. The codes for each digit follow the binary sequence, starting with $0_{10} = 0000_2$, $1_{10} = 0001_2$, $2_{10} = 0010_2$ through to nine, which is represented by $1001_2 = (1 \times 8 + 0 \times 4 + 0 \times 2 + 0 \times 1)_{10}$. Note that four bits are needed for each BCD digit since this allows coding of the ten digits. Six of the codes possible for the four bits (1010_2 through to 1111_2) are unused in BCD. This system has been popular in calculators.

This system is sometimes called 8421 BCD due to the natural binary weights used. 2421 BCD weights the three least significant digits using natural binary, but the most significant digit is only weighted as two, not eight as is 8421 BCD, and the maximum total weight is 9 as in the decimal system.

Octal is a decimal coded version of a radix 8 binary system. According to the radix, eight digits are used and these are 0, 1, 2, 3, 4, 5, 6, 7, the first

eight decimal symbols. Each digit represents a collection of three binary digits, which are coded using decimal symbols. There is a type definition for octal in the C++ programming language, but the system is less used now though it found early popularity in minicomputer systems.

Table 7.1 Different codes

Decimal$_{10}$	Binary$_2$	BCD	Octal$_8$	Hexadecimal$_{16}$
0	0000	0000	0	0
1	0001	0001	1	1
2	0010	0010	2	2
3	0011	0011	3	3
4	0100	0100	4	4
5	0101	0101	5	5
6	0110	0110	6	6
7	0111	0111	7	7
8	1000	1000	10	8
9	1001	1001	11	9
10	1010	0001 0000	12	A
11	1011	0001 0001	13	B
12	1100	0001 0010	14	C
13	1101	0001 0011	15	D
14	1110	0001 0100	16	E
15	1111	0001 0101	17	F
16	10000	0001 0110	20	10
17	10001	0001 0111	21	11
18	10010	0001 1000	22	12

Hexadecimal is a widely used system of radix 16 employing encoded binary numbers. Since binary numbers represent a large collection of bits, it is often convenient to compact them using a code. The hexadecimal system encodes four bit nibbles and, since a pattern of four bits has 16 possible values, we need symbols for 16 digits. Decimal symbols are used to represent the first nine of these, so $0_{16} = 0000_2$, $1_{16} = 0010_2$, $2_{16} = 0010_2 \ldots 9_{16} = 1001_2$. The first six letters of the alphabet are used to represent the remaining six combinations: $A_{16} = 1010_2$ (10_{10}); $B_{16} = 1011_2$ (11_{10}); $C_{16} = 1100_2$ (12_{10}); $D_{16} = 1101_2$ (13_{10}); $E_{16} = 1110_2$ (14_{10}); and $F_{16} = 1111_2$ (15_{10}). Hexadecimal is popular throughout computer systems that often use eight-bit bytes or 16-bit or 32-bit words. These can conveniently be packaged as collections of four-bit nibbles represented

using the hexadecimal system. A count sequence illustrates the relationship between these different number systems.

Table 7.1 illustrates that a more compact number representation is achieved with a high radix, though this requires more symbols for the digits.

> *The Assyrian radix was 60 and it still remains in our division of time and angles. One plausible reason for the decimal system is that the number of our fingers and thumbs is ten. Don't infer too much about the Assyrian physiology!*

7.1.2 *Conversion between systems*

The most important code conversion step is from decimal to binary. This is because the other systems are essentially binary based. In order to convert a decimal **integer** to binary, one algorithm is to divide a number by two and the remainder left at each stage becomes a bit in the converted binary number. The remainder from the first division is the least significant bit (LSB) of the resulting number and bits are generated in increasing order of significance. The number resulting from division is then divided by two, and the remainder from division becomes the bit of next most significant precision. This process continues successively, until the result of division is one, the most significant bit (MSB) of the resulting number. Consider 9_{10} and 23_{10} converted into binary:

$9/2 = 4$ remainder 1 LSB	$23/2 = 11$ remainder 1 LSB
$4/2 = 2$ remainder 0	$11/2 = 5$ remainder 1
$2/2 = 1$ remainder 0	$5/2 = 2$ remainder 1
$1/2 = 0$ remainder 1 MSB	$2/2 = 1$ remainder 0
	$1/2 = 0$ remainder 1 MSB
$9_{10} = 1001_2$	$23_{10} = 10111_2$

In order to convert a binary integer to decimal, one method is to multiply the MSB by two and then add the bit of next least significance. This continues successively until the LSB is added. For 10111_2, for which (as just

illustrated) $10111_2 = 23_{10}$, the process is:

10111_2	10111_2
$1 \times 2 + 0$	$1 \times 2 + 0 = 2$
$(1 \times 2 + 0) \times 2 + 1$	$2 \times 2 + 1 = 5$
$((1 \times 2 + 0) \times 2 + 1) \times 2 + 1$	$5 \times 2 + 1 = 11$
$(((1 \times 2 + 0) \times 2 + 1) \times 2 + 1) + 1$	$11 \times 2 + 1 = 23_{10}$

and there are similar algorithmic techniques to convert **fractions**. Conversion between octal and binary and between hexadecimal and binary is much easier, since octal and hexadecimal are coded versions of binary. If we take a 12-bit word we can just partition it into three four-bit sections for hexadecimal, or into four three-bit sections for octal, e.g.

Bit	1	2	3	4	5	6	7	8	9	10	11	12
Binary	0	1	0	0	1	1	0	1	0	1	0	1
Octal	2			3			2			5		
Hexadecimal	4				D				5			

$$010011010101_2 = 2325_8 = 4D5_{16}$$

7.1.3 *Other codes*

Morse code is historically of great importance since it was first used in telegraphic transmission. It is not used in digital circuits since the symbols vary in length. The American standard code for information interchange, *ASCII*, is a seven-bit alphanumeric code used in many computer systems and is the code used to represent characters, as is the extended binary-coded decimal interchange code (*EBDIC*). A code of major importance is the Gray code, which has already been introduced in ASM (algorithmic state machine) chart design (Chapter 6). This code is a minimal distance code since one bit only changes between elements of the code. It was originally used in rotational transducers to avoid the problems with alignment that a natural binary code would incur. *Excess 3* is a code equivalent to the BCD coding plus three, hence the name. This implies that the symbols use the

binary codes for 3 to 12 inclusive and the codes for 0, 1, 2 and 13, 14, 15 are not used.

Decimal	BCD	Excess 3
0	0000	0011
1	0001	0100
2	0010	0101
3	0011	0110
4	0100	0111
5	0101	1000
6	0110	1001
7	0111	1010
8	1000	1011
9	1001	1100

7.2 Binary Arithmetic

7.2.1 *Basic binary arithmetic*

Basic arithmetic naturally concentrates on the main arithmetic functions, addition, subtraction, multiplication, and division. We shall consider **unsigned** numbers first and, to simplify presentation, we shall consider positive integers only.

7.2.1.1 *Binary addition*

When adding two binary digits we generate a *sum* bit and a *carry* bit. The sum is the result of addition at that stage and the carry is generated to be included in the addition at the stage of next most significance. The rules for binary, modulo 2, addition of two bits A and B, $A + B$, are:

Bit 1		Bit 2		Sum	Carry
0	+	0	=	0	0
0	+	1	=	1	0
1	+	0	=	1	0
1	+	1	=	0	1

When adding two binary numbers, say 110_2 (6_{10}) and 101_2 (5_{10}), these rules are used:

$$
\begin{array}{cccc}
1 & 1 & 0 & \\
1 & 0 & 1 & + \\
\hline
0 \quad 0 & 1 & 1 & \text{Sum bit} \\
1 & 0 & 0 & \text{Carry bit} \\
\hline
1 \quad 0 & 1 & 1 & \text{Result of addition}
\end{array}
$$

The result of adding these two numbers is then 1011_2, which is 11_{10} as expected. These rules can be implemented in **hardware** and we can design circuitry that implements binary addition. If the numbers are presented in parallel, then we need to design circuits that add two bits at a time and generate a carry to the next stage. The general format is shown in Figure 7.1.

Figure 7.1 Addition of two binary numbers

The addition rules provide a way to add two bits to provide a sum bit and a carry bit. The rules actually describe a truth table where the sum bit is '1' when either bit is '1' and is then the EXCLUSIVE OR of the two bits added. The carry bit is set only when both bits are '1' and is the AND of the two input bits. The circuit implementing the rules is known as a *half adder*. This circuit can be used to add the two LSBs of a binary number. The rules for **full** addition describe the generation of a sum and carry bit for two input bits plus the carry generated from a previous stage. This is Bit 0 in the addition circuit in Figure 7.1. If we label Number 1 Bit 0 as $N1B_0$,

and $N2B_0 =$ Number 2 Bit 0, the rules for addition are:

$N1B_0 + N2B_0$		Sum	Carry-out
0	0	0	0
0	1	1	0
1	0	1	0
1	1	0	1

This is the definition of a combinational logic circuit. The sum bit is the XOR function and the carry-out is AND and the circuit is shown in Figure 7.2.

(a) Symbol (b) Circuit

Figure 7.2 Half adder

If the bits to be added are A_i and B_i, generating a sum bit S_i and a carry bit C_i, given the carry generated by the previous stage, the bit of next least significance is C_{i-1}. The rules for addition are then:

C_{i-1}	+	A_i	+	B_i	S_i	C
0		0		0	0	0
0		0		1	1	0
0		1		0	1	0
0		1		1	0	1
1		0		0	1	0
1		0		1	0	1
1		1		0	0	1
1		1		1	1	1

The carry input at each stage is often known as the carry-in. The generated carry is often known as the carry-out, but don't confuse it with a pizza or a six-pack!

A *full adder* is a circuit that implements these rules. In order to design a minimised circuit for the sum bit S_i we can draw up its K-map:

$C_{i-1}\backslash A_i B_i$	00	01	11	10
0	0	1	0	1
1	1	0	1	0

The K-map does not give much help for minimisation in this case since we cannot loop any common terms. By extracting each prime implicant alone we get

$$S_i = \overline{C}_{i-1} \cdot \overline{A}_i \cdot B_i + \overline{C}_{i-1} \cdot A_i \cdot \overline{B}_i + C_{i-1} \cdot \overline{A}_i \cdot \overline{B}_i + C_{i-1} \cdot A_i \cdot B_i$$

which by grouping terms can achieve an implementation using XOR gates only:

$$S_i = \overline{C}_{i-1} \cdot (\overline{A}_i \cdot B_i + A_i \cdot \overline{B}_i) + C_{i-1} \cdot (\overline{A}_i \cdot \overline{B}_i + A_i \cdot B_i)$$
$$S_i = \overline{C}_{i-1} \cdot (A_i \oplus B_i) + C_{i-1} \cdot (\overline{A_i \oplus B_i}) = A_i \oplus B_i \oplus C_{i-1}$$

The carry bit C_i is '1' when two or more of the input bits are '1' so

$$C_i = C_{i-1} \cdot A_i + C_{i-1} \cdot B_i + A_i \cdot B_i$$

The circuit for a full adder is shown in Figure 7.3.

Figure 7.3 Circuit for full adder

The circuit to add two three-bit binary numbers is shown in Figure 7.4. This circuit uses full adders only, so the carry-in to the least significant addition is set to '0'. This can be compared with the circuit of Figure 7.1.

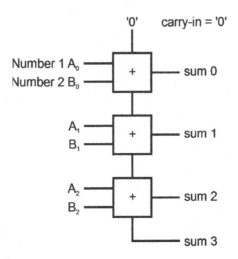

Figure 7.4 Addition of two three-bit binary numbers

We can then generate the sum and carry outputs using standard logic gates. By further analysis of the carry signal we obtain

$$C_i = C_{i-1} \cdot (\overline{A_i} \cdot B_i + A_i \cdot \overline{B_i} + A_i \cdot B_i) + A_i \cdot B_i$$
$$= C_{i-1} \cdot (\overline{A_i} \cdot B_i + A_i \cdot \overline{B_i} + A_i \cdot B_i) + A_i \cdot B_i$$
$$= C_{i-1} \cdot (\overline{A_i} \cdot B_i + A_i \cdot \overline{B_i}) + A_i \cdot B_i$$
$$= C_{i-1} \cdot (A_i \oplus B_i) + A_i \cdot B_i$$

and so a full adder can be implemented using two half adders and an OR gate. By denoting a half adder as HA, the circuit to add two binary numbers is as per Figure 7.5.

Figure 7.5 Addition via half adders

Arithmetic circuits have been a very fertile area of logic circuit development due to the importance of addition in computers. Carry look-ahead circuits can improve speed by testing the carry for all stages simultaneously, rather than the circuit provided where the carry propagates from one stage to the next, reducing potential speed. Now that memory is cheap, why not just use look-up tables?

7.2.1.2 Binary subtraction

When subtracting two binary digits we generate a *difference* bit which when '1' indicates that one bit is '1' and the other is '0' (hence indicating that there is a difference between them), and a *borrow* bit. The borrow is generated to indicate that the number subtracted is greater in size than the number it is being subtracted from (i.e. a negative result would ensue) and when subtracting bit B from bit A the borrow bit is '1' when A is '0' and B is '1'. Formally, the *minuend* is subtracted from the *subtrahend* to provide the difference between the two numbers. For two bits, to subtract B from A (to form A-B), the rules and circuit are given in Figure 7.6.

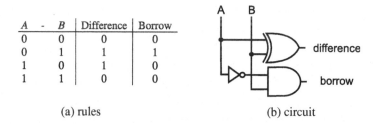

A	-	B	Difference	Borrow
0		0	0	0
0		1	1	1
1		0	1	0
1		1	0	0

(a) rules (b) circuit

Figure 7.6 Half subtracter

These rules define a *half subtracter* where the difference is identical to the sum bit for a half adder. The half subtracter is used for the LSBs and a *full subtracter* (Figure 7.7) is used for the remainder of the bits, to subtract B_i from A_i given a borrow Br_{i-1} to generate a difference D_i and borrow Br_i.

We could then use the circuits for the full and half subtracters to provide a full circuit to subtract two binary numbers. This is usually avoided in

(a) rules (b) circuit

Figure 7.7 Full subtracter

signed binary arithmetic by using a complement form that employs addition only and so we do not need subtracter circuits. We shall complete the basic binary arithmetic systems before looking at signed binary arithmetic.

7.2.1.3 *Binary multiplication*

In principle, binary multiplication is straightforward since anything multiplied by zero results in zero and anything multiplied by one remains the same. The rules for multiplication are then:

A	B	$A \times B$
0	0	0
0	1	0
1	0	0
1	1	1

The procedure for binary multiplication is similar to that often used for decimal multiplication. Formally, the number multiplied is called the *multiplicand*, and the number multiplying it is called the *multiplier*.

In decimal arithmetic we introduce zeros corresponding to the position of the most significant digit and then multiply the multiplicand by that digit. On the next row we introduce zeros corresponding to the next most significant digit and then multiply by that digit. This continues to the row corresponding to the least significant digit where we introduce no zeros and insert the result of multiplication. We then add up the results for each digit

in to form the result. The results at each intermediate step are the *partial products*. For example, 53_{10} multiplied by 321_{10}:

Multiplicand			5	3		
Multiplier			3	2	1	\times
(300×53)	1	5	9	0	0	Partial product for most significant digit 300
(20×53)		1	0	6	0	Partial product for 20
(1×53)				5	3 $+$	Partial product for least significant digit 1
	1	7	0	1	3 $=$	Result

The process of introducing zeros is equivalent to multiplying the multiplicand by a factor of ten, which, by virtue of the decimal system, is equivalent to shifting the number left and introducing zeros into the vacated digits and this can be seen in the partial products. The same procedure is used in binary multiplication; the multiplicand is shifted left according to the position or significance of the bit in the multiplier. The results are all summed to provide the final result and so the process is called *shift and add*. Each partial product is formed by multiplying a binary number by a factor of two, which is equivalent to shifting the number left and adding zeros to the vacated positions. For multiplication by four, we shift the number left two places, e.g. $1011_2 \times 100_2$ (2^2).

$$1\ 0\ 1\ 1 \times 1\ 0\ 0 = 1\ 0\ 1\ 1\ 0\ 0$$

This is used to form each partial product when multiplying by a binary integer, e.g. 1011_2 multiplied by 1101_2.

Multiplicand				1	0	1	1
Multiplier				1	1	0	1 \times
(1000×1011)	1	0	1	1	0	0	0
(100×1011)		1	0	1	1	0	0
(1×1011)				1	0	1	1 $+$
	1	0	0	0	1	1	1 $=$

The resulting binary number is then 143_{10} (the multiplier is 13_{10} and the multiplicand is 11_{10}). One partial product is omitted in the summation. This is because one bit is 0 in the multiplier (corresponding to 2^1).

The shift-and-add technique lends itself readily to a hardware imple-
mentation. Since multiplying by 2^n is equivalent to shifting left by n places,
this multiplication can then be implemented using a shift register. The
outputs of the shift register are gated from another shift register holding
the multiplier to give an input to an adder circuit. The multiplier is stored
in a shift register to facilitate testing whether bits are set or not. The other
number input to the adder is from the outputs of a register that holds the
final result of multiplying the two numbers, formed by accumulating the
partial products. See Figure 7.8.

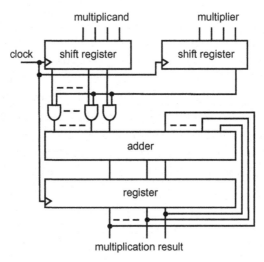

Figure 7.8 Multiplication circuit

At the first clock cycle the multiplicand is fed to the adder only if the
multiplier LSB is '1'; otherwise the first partial product is zero. The first
partial product is then stored in the register. At the next clock cycle the
multiplicand is shifted left one place (multiplying it by two) and if the
next MSB in the multiplier is '1' then this second partial product is added
to the first. The register then contains the result of adding the first two
partial products. Multiplication continues in this manner by shifting the
multiplicand to form the partial products and adding it to the contents of
the register according to whether appropriate bits are set in the multiplier.
The process terminates when the partial product resulting from the MSB is
added to the final sum. The register then contains the result of multiplication.
The register needs to be sufficiently large to accommodate the result and in

this case the size should at least be the joint length of the multiplier and the multiplicand. In a software version, the (pseudocode) procedure to multiply `multiplicand` by `multiplier` is:

```
result := 0 ;initialise the storage of the summation
FOR i=1 TO number_of_bits DO
    BEGIN
    IF (multiplier[i]='1') result:= result (plus)
    multiplicand
    ;If the bit is set, add the partial product to the result
    multiplicand=multiplicand*2
    ;Now shift the multiplicand left to form the next partial
    product
    END
```

Multiplication of binary integers is easily accomplished using a *look-up table*, the implementation is discussed later in Section 7.3.1.

7.2.1.4 *Binary division*

Binary division is more complex. It can use a procedure equivalent to that for conventional longhand division. Formally, the number divided is known as the *dividend*, the number it is divided by is the *divisor*, and the result is the *quotient*. Note that we can form a result with a fractional part. At each stage we determine how many times the divisor is an integer number of divisions of the dividend. We start with the most significant number of divisions and work towards the bits of least significance. At each stage we perform binary subtraction and bring down the next least significant digit from the dividend. We can add trailing zeros to the dividend to increase the number of digits in a fractional result. For example, 1110_2 divided by 100_2 (the decimal equivalent is shown on the right):

		1	1.	1_2					3.	5_{10}
100_2	1	1	1	0.	0_2	4_{10}	14.	0_{10}		
	1	0	0					12		
		1	1	0				2	0	
		1	0	0				2	0	
			1	0	0			0	0	
			1	0	0					
			0	0	0					

The result of dividing 14_{10} by 4_{10} is then 3.5_{10} as expected and in this case the procedure terminated with an exact fraction. This is not usually the case where division continues until a specified number of bits have been generated. This procedure is highly complex to implement as a digital system.

Note that you can store the representation of the reciprocal of a number in memory. Division then becomes a process of multiplication by the reciprocal of a number, rather than dividing by the number, and this is much faster.

7.2.2 Signed binary arithmetic

7.2.2.1 Signed arithmetic

In order to handle signed binary numbers we could use an extra bit, a *sign bit*, and set it to '0' for positive numbers or '1' for a negative number. When adding or subtracting two numbers we could use the sign bit to choose the output from an adder or a subtracter, respectively. This implies that we need different circuits to add and subtract numbers and, as shown earlier, though the sum bit is the same as the difference, the borrow is completely different from the carry. Using separate adders and subtracters would therefore increase cost. It is better to use *complement arithmetic* where we represent a number in its complement, or negative form, and use addition only. Instead of performing a subtraction $A - B$, we form the complement of B, $-B$, and add this to A to form the subtraction as $A - B = A + (-B)$. Though we then need the means to generate the complement form of a number, we only need addition units and not those for subtraction as well. This is much cheaper than using adders and subtracters in tandem.

There are two ways to generate the complement; there is a **radix** complement and a **radix-1** complement. In decimal arithmetic there is a *ten's complement* and a *nine's complement*; in binary there is a *two's complement* and a *one's complement* for the radix and radix-1 complement, respectively. For an n-digit number N, the radix complement is $r^{n+1} - N$ and the radix-1 complement is $(r^{n+1} - 1) - N$. For an n-digit

number N:

Ten's complement $10^{n+1} - N = 10000..00 - N$

Nine's complement $10^{n+1} - 1 - N = 10000..00 - 1 - N = 9999..99 - N$

Two's complement $2^{n+1} - N = 1000..00 - N$

One's complement $2^{n+1} - 1 - N = 1000..00 - 1 - N = 111..11 - N$

When the number is in complement form we have no way of knowing this. We therefore include a designated sign bit, which in binary will be '1' to indicate a negative number and '0' a positive one. Note that we include the sign bit as '0' prior to complementing a number. When the number is in complement form the sign bit will be '1'. Having formed the complement we then need to perform addition only and have no need for extra subtraction circuitry.

Note that Excess 3 is sometimes termed 'self-complementing' due to the cyclic nature of the complements of elements of the code. The nine's complement of 3 is 6 and that of 9 is 1.

As well as including an extra bit to indicate the sign of a number we can include a further *overflow* bit to handle a situation where the result might be too large to fit within a specified bit pattern. For example, if we add two four-bit numbers 1000_2 (8_{10}) and 1001_2 (9_{10}), we generate a result 10001_2 (17_{10}), which is a five-bit number with the fifth bit generated by the size of the two four-bit numbers. If we did not specify an extra bit to accommodate the overflow then this number could be interpreted to be a negative four-bit number (since the sign bit is '1'), a confusing result. To add these numbers, the fifth bit is the overflow bit and the sixth bit is the sign bit. The overflow bit will not be included in examples here but needs to be borne in mind by a designer.

7.2.2.2 *One's complement arithmetic*

In order to generate the one's complement we need to subtract the number from a set of '1's. For a three-bit number $B_2 B_1 B_0$, we first include an extra

bit as the sign bit, the (positive) number then becomes $0B_2B_1B_0$, and we then subtract the number from a set of '1's to form the complement:

$$
\begin{array}{ccccl}
 & 1 & 1 & 1 & 1 \\
- & 0 & B_2 & B_1 & B_0 & \text{(positive) original number} \\
\hline
= & 1 & \overline{B_2} & \overline{B_1} & \overline{B_0} & \text{(negative) number in one's complement form}
\end{array}
$$

This shows that **to form the one's complement** we merely need to **invert each bit**. The sign bit becomes '1' indicating that the number is in complement form. We can now add to perform subtraction. For example, let us consider $2_{10} - 8_{10}$ where $2_{10} = 0010_2$ and $8_{10} = 1000_2$. We first add a sign bit to each number, then complement 8_{10} and then add the complemented form of 8_{10} to 2_{10}.

$$8_{10} = 1000_2 \text{ so adding a sign bit } 8_{10} = 01000_2 \text{ and}$$
$$\text{the one's complement} - 8_{10} = 10111_2$$

To perform $2_{10} - 8_{10}$

$$
\begin{array}{lr}
+2 & 00010 \\
-8 & 10111 \\
\hline
\text{Addition} & 11001
\end{array}
$$

By inspecting the result we can see that the sign bit is '1'. This implies that the resulting number is negative and in one's complement form. To find out what the number is we merely need to invert each bit to see that it represents -00110 or -6, the expected result.

The number has only been inverted here to illustrate that the correct result has been achieved. When complement arithmetic is used in arithmetic in computers, or in other digital systems, the complement is generated only at the input or at the output stage. At the input stage we can feed negative numbers to a digital system by generating their complement and at the output stage we can invert the complement to see the number the complement represents. For the remainder of the arithmetic process the number remains in its complement form, we do not need to generate complements during arithmetic, only before or after.

Consider also $8_{10} - 2_{10}$. Given that $8_{10} = 1000_2$ and $2_{10} = 0010_2$. We need to add a sign bit to each number and complement 2_{10} then add the complemented form to 8_{10}.

$$2_{10} = 0010 \text{ plus sign bit } 2_{10} = 00010 \text{ and}$$
$$\text{the one's complement} - 2_{10} = 11101$$

To perform $2_{10} - 8_{10}$	+8	01000	take number
	−2	11101	add complement
	Addition	100101	

When we inspect the sign bit we can see that it represents a positive number since the sign bit, bit four is '0'. Our main problem is that there is an extra bit above the sign bit and the other difficulty is that the answer appears to be wrong! The result of $8 - 2 = 5$! This is actually inherent in one's complement arithmetic since we have performed $8 + 2^{n+1} - 1 - 2 = 5 + 2^{n+1}$, the given result. All the extra bit tells us is that the number is out by one by virtue of one's complement arithmetic. We then add the extra '1' (if it is generated) back in to the LSB in a process known as *end-around-carry*. E.g. $14_{10} - 3_{10}$:

3	00011	
−3	11100	
14	01110	
14−3	101010	
end-around-carry		1 +
result	01011	=

When the end-around-carry is implemented, the result is $01011_2 = 11_{10}$, as expected. Since the end-around-carry is inherent in one's complement arithmetic it makes it a **two-stage process**. There are two addition stages; one is to correct the result. By virtue of this, two's complement arithmetic is universally preferred since it does not need this extra correction stage. Also, there are two representations of zero in one's complement arithmetic, both 0000..0 and 1111..1 represent zero (the latter in one's complement form), and this can lead to extra difficulty.

7.2.2.3 *Two's complement arithmetic*

The two's complement of an n-bit number N is defined as $2^{n+1} - N$. This is the one's complement plus one. The one's complement was generated by inverting each digit, so **to form the two's complement** we need to **invert**

each bit then add one. We need to include an extra sign bit and after the complementing process the sign bit will be '1' indicating a negative number in two's complement form. Again, having formed the complement we then need to perform addition only and we do not need subtraction. E.g. $2_{10} - 8_{10}$:

$$8_{10} = 1000 \quad \text{add sign bit,} \quad +8_{10} = 01000,$$

two's complement = invert each bit and add one so

$$-8_{10} = 10111 + 1 = 11000_2$$

$$
\begin{array}{ll}
+2_{10} & 00010 \\
-8_{10} & 11000 \\
\hline
2 + (-8) \quad = & 11010
\end{array}
$$

If the subtraction is to be achieved by two's complement addition implemented using full adders, then we invert each bit of the number to be complemented and set C_{in} for the LSB to '1'. See the design exercises in Section 7.3.

When we inspect the result, the sign bit is '1' indicating a negative number in two's complement form. To determine the negative number it represents we merely repeat the complementing process, invert each bit, and add one:

$$11010 = -00101 + 1 = -00110 = -6_{10}$$

This shows that the result is as expected. Note again that once the number is in complement form, it remains so for the remainder of the arithmetic process. As with one's complement arithmetic we are only inverting the complemented number to show that the correct result has been achieved.

Finally, let us consider $8_{10} - 2_{10}$:

$$
\begin{array}{lll}
+2_{10} & 00010 & \\
-2_{10} & 11101 & + \quad 1 \\
\hline
-2_{10} & 11110 & \\
+8_{10} & 01000 & + \\
\hline
& 100110 & =
\end{array}
$$

If we ignore the extra bit above the sign bit, the result is 00110, $+6_{10}$, as expected. The extra bit can actually be ignored since it is inherent in two's

complement arithmetic. We have performed $8 + 2^{n+1} - 2 = 6 + 2^{n+1}$. The extra bit corresponds to 2^{n+1}, which was only introduced as part of the complementing process and can therefore be ignored. Two's complement arithmetic has **one stage only**, as opposed to one's complement, which requires two stages. For this reason two's complement arithmetic is universally preferred. Also, there is a single representation of zero, 0000..0, in two's complement arithmetic by virtue of the complementing process.

7.2.3 *Fixed- and floating-point arithmetic*

General-purpose arithmetic systems need to handle numbers with both fractional and integer parts. Two representations exist: *floating-point* and *fixed-point* arithmetic. Floating-point arithmetic represents a number as a *mantissa* and *exponent*. The mantissa is a binary fraction expressed in two's complement form. The exponent is binary and again expressed in complement form. The floating-point representation is then

$$\text{number} = \text{mantissa} \times \text{exponent} = m \times 2^e$$

where m, the mantissa, is a fractional binary number, $m = \sum d_i 2^{-i}$ where $i \in 1, n$ for an n-bit mantissa, and e is the exponent. This is expressed as a pattern of bits, usually stored within a computer word. Floating-point systems usually employ a specified format, amongst which the IEEE/ANSI 754-2008 is often preferred in computer systems. Fixed-point systems are usually application-specific for reasons of speed and use a floating-point number system arranged explicitly for the application and generally without using an exponent as in floating-point systems.

An arrangement of a word for a floating-point number is then:

no. of bits	1	n	1	p
contents	Exponent sign	n-bit mantissa	Mantissa sign	p-bit exponent

The mantissa is usually normalised so that $0.5 \leq \text{mantissa} < 1.0$. This is consistent with the binary exponent and since the first bit of the number is 1 it ensures that all the available *resolution* is used. The resolution describes how well the binary number approximates its intended value. The resolution is limited since we are only storing a limited, and not infinite, set of bits to

represent a number. Common arithmetic systems in computers use 16-bit and 32-bit arithmetic and some languages allow the use of extra *precision* to store numbers to expedite better accuracy. For example, in a system with a six-bit mantissa and a three-bit exponent:

$$1.6875_{10} = 0.84375 \times 2^1$$
$$= (1 \times 0.5 + 1 \times 0.25 + 0 \times 0.125$$
$$+ 1 \times 0.0625 + 1 \times 0.03125) \times 2^1$$
$$= 0.11011\text{E}001$$

The normalisation process reduces the number to be within the correct range and the exponent is increased to reflect by how much the mantissa has been reduced. The slash between the two binary numbers indicates the partition between the mantissa and exponent. If the normalised number has been increased to fit within the range then the exponent will be negative, indicating by how much the mantissa needs to be reduced to return to the original number. The negative exponent will be stored in two's complement form; for example, 0.3128_{10} needs to be multiplied by four to fit within the normalisation range so

$$0.1328_{10} = 0.53125 \times 2^{-2}$$
$$= (1 \times 0.5 + 0 \times 0.25 + 0 \times 0.125$$
$$+ 0 \times 0.0625 + 1 \times 0.03125) \times 2^{-2}$$
$$= 0.1001\text{E}110$$

where the leading '1' in the exponent is a sign bit indicating a number expressed in two's complement form representing -2. The size of the exponent restricts it to be between -4 (100) and $+3$ (011).

In order to add two floating-point numbers they first need to be denormalised so that the exponents are equal. The mantissas are then of the same range and they can be added. The number with the smallest exponent is denormalised by dividing it by two and then increasing the exponent by one. This double process continues until the exponent equals that of the other number when the two mantissas represent fractions in the same range. For example, $1.6875_{10} + 0.1328_{10}$:

$1.6875_{10} = 0.11011\text{E}001$, this has the largest of the two exponents and

is thus the target for normalisation.

$0.1328_{10} = 0.10001E110$ so shift the mantissa right and add
 one to the exponent

$= 0.01000E111$ the exponent is now -1 so continue

$= 0.00100E000$ the exponent is now zero so we need
 one more step

$= 0.00010E001$ the final denormalised number

	1.6875_{10}		$0.11011E001$
$+$	0.1328_{10}	$+$	$0.00010E001$
$=$	1.8203_{10}	$=$	$0.11101E001$

The result of binary floating-point addition actually represents 1.8125_{10}, which does not equal the result of decimal arithmetic due to the decreased resolution of the five-bit mantissa in the floating-point representation. Information was lost during denormalisation of 0.1328_{10} (the final '1' in the normalised number was lost during denormalisation since it did not fit within a five-bit mantissa). This is the effect of limited precision in the arithmetic process.

The processes of multiplication and division are given mathematically for floating point numbers by:

$$\text{Multiplication: } (A \times 2^a) \times (B \times 2^b) = (A \times B) \times 2^{(a+b)}$$

$$\text{Division: } (A \times 2^a)/(B \times 2^b) = (A/B) \times 2^{(a-b)}$$

These processes might require renormalisation of the resulting mantissa to ensure that it remains within the normalisation range.

7.3 Design Examples

A number of design examples are included to illustrate not only how arithmetic systems can be implemented, in particular within programmable combinational circuits, but also be used for design purposes to facilitate implementation. Arithmetic circuits are of natural interest to computer designers and there has been much research directed at their implementation. The presentation here is to clarify the main academic material whilst providing pointers to the issues that system designers need to resolve.

7.3.1 *Binary arithmetic*

The first design shows the duality between mathematics and combinational design. We shall consider the design of a circuit to multiply two two-bit positive binary integers. This can be implemented as a combinational logic design since the product of the two numbers is fixed. The combinational logic design can be specified using a truth table with design based on a K-map. The K-map is a rather cumbersome approach though it is good exercise in combinational design. To multiply a binary number $A_1 A_0$ by $B_1 B_0$ to produce a result $M_3 M_2 M_1 M_0$ where A_1, B_1, and M_3 are the MSBs, the truth table is:

A_1	A_0	×	B_1	B_0	M_3	M_2	M_1	M_0
0	0	X	X		0	0	0	0
X	X	0	0		0	0	0	0
0	1	0	1		0	0	0	1
0	1	1	0		0	0	1	0
0	1	1	1		0	0	1	1

A_1	A_0	×	B_1	B_0	M_3	M_2	M_1	M_0
1	0	0	1		0	0	1	0
1	0	1	0		0	1	0	0
1	0	1	1		0	1	1	0
1	1	0	1		0	0	1	1
1	1	1	0		0	1	1	0
1	1	1	1		1	0	0	1

By K-map analysis,

$$M_0 = A_0 \cdot B_0$$
$$M_1 = \overline{A_1} \cdot A_0 \cdot B_1 + A_1 \cdot \overline{A_0} \cdot B_0 + A_1 \cdot \overline{B_1} \cdot B_0 + A_0 \cdot B_1 \cdot \overline{B_0}$$
$$M_2 = A_1 \cdot \overline{A_0} \cdot B_1 + A_1 \cdot B_1 \cdot \overline{B_0}$$
$$M_3 = A_1 \cdot A_0 \cdot B_1 \cdot B_0$$

Another method of designing the circuit is to use shift and add as the design process. The multiplication result is the sum of $B_1 B_0$ shifted right one place, which is added to a sum if A_1 is set. $B_1 B_0$ is added to the sum if A_0 is set. The design can be achieved by evaluating the sum bit at each stage given the carry generated by the previous stage:

		B_1	B_0		
		A_1	A_0	×	
$A_1 \cdot B_1$	$A_1 \cdot B_0$	0			
	$A_0 \cdot B_1$	$A_0 \cdot B_0$		+	
M_3	M_2	M_1	M_0		=

For the first bit, $M_0 = 0$ plus $A_0 \cdot B_0 = A_0 \cdot B_0$
which generates no carry

For the second bit, the addition (Section 7.2.1.1) — the sum — is an XOR function

$$M_1 = A_1 \cdot B_0 \text{ plus } A_0 \cdot B_1 = \overline{A_1 \cdot B_0} \cdot A_0 \cdot B_1 + A_1 \cdot B_0 \cdot \overline{A_0 \cdot B_1}$$

by De Morgan's law $M_1 = (\overline{A_1} + \overline{B_0}) \cdot A_0 \cdot B_1 + A_1 \cdot B_0 \cdot (\overline{A_0} + \overline{B_1})$

and the carry is AND, so $\text{Carry}_1 = A_1 \cdot B_0 \cdot A_0 \cdot B_1$

For the third bit $M_2 = A_1 \cdot B_1$ plus $Carry_1$

So $M_2 = A_1 \cdot B_1 \cdot \overline{A_1 \cdot B_0 \cdot A_0 \cdot B_1} + \overline{A_1 \cdot B_1} \cdot A_1 \cdot B_0 \cdot A_0 \cdot B_1$

By De Morgan's law $M_2 = A_1 \cdot B_1 \cdot (\overline{A_1} + \overline{B_0} + \overline{A_0} + \overline{B_1}) + (\overline{A_1} + \overline{B_1})$
$\cdot A_1 \cdot B_0 \cdot A_0 \cdot B_1$

So $M_2 = A_1 \cdot B_1 \cdot \overline{B_0} + A_1 \cdot B_1 \cdot \overline{A_0}$

this generates a carry $Carry_2 = A_1 \cdot B_0 \cdot A_0 \cdot B_1$

$M_3 = $ the carry from the previous stage $M_3 = A_1 \cdot B_0 \cdot A_0 \cdot B_1$

These equations for M_3, M_2, M_1, and M_0 are the same as those by combinational design, merely derived by mathematical analysis as well as by K-maps, illustrating the duality between the two techniques.

7.3.2 Code conversion

7.3.2.1 Binary-to-BCD conversion

We shall now design a system to convert from four-bit binary to BCD, the codes shown in Section 7.1.1. The conversion we want to achieve is:

Binary				BCD					Binary				BCD				
B_3	B_2	B_1	B_0	D_4	D_3	D_2	D_1	D_0	B_3	B_2	B_1	B_0	D_4	D_3	D_2	D_1	D_0
0	0	0	0	0	0	0	0	0	1	0	0	0	0	1	0	0	0
0	0	0	1	0	0	0	0	1	1	0	0	1	0	1	0	0	1
0	0	1	0	0	0	0	1	0	1	0	1	0	1	0	0	0	0
0	1	0	0	0	0	1	0	0	1	1	0	0	1	0	0	0	1
0	1	0	1	0	0	1	0	1	1	1	0	1	1	0	0	1	0
0	1	1	0	0	0	1	1	0	1	1	1	0	1	0	1	0	0
0	1	1	1	0	0	1	1	1	1	1	1	1	1	0	1	0	1

We wish to design a system that, given the four-bit binary input, will produce the specified five-bit binary output. There are a number of ways to achieve the design and we could use:

(i) Memory.
(ii) Combinational design via K-maps.
(iii) Combinational design via mathematics.

The memory implementation would require specifying the required conversion table as the contents of *a read-only memory* (ROM). Each of the five outputs has 16 slots addressed by the four input bits. This requires a ROM with 80 locations only. This is a comparatively expensive solution. Cost can be reduced by using either a programmable array logic (PAL), or in a chip design by reducing the gate count. The design is therefore performed first by traditional K-map design which is compared with the result of a PAL implementation. K-maps are really limited to small-scale problems only so we shall also use two's complement arithmetic for minimisation. This technique is much better suited to circuits with more inputs and that can be described using mathematical techniques.

The minimisation techniques should all achieve the same (or an equivalent) result, which they do. They reach this solution by different ways, each highlighting the benefits and practice of a chosen design technique.

7.3.2.2 *Combinational design*

By inspection of the code conversion to be implemented we can see that D_0 is a direct copy of B_0. So the output D_0 is given by $D_0 = B_0$. By K-map analysis for D_1 and D_2:

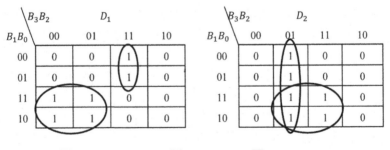

so $D_1 = \overline{B}_3 \cdot B_1 + B_3 \cdot B_2 \cdot \overline{B}_1$ and $D_2 = \overline{B}_3 \cdot B_2 + B_2 \cdot B_1$

D_3 is set only in two cases so minimisation does not need a K-map. D_4 is '1' only when B_3 is '1' and B_2 or $B_1 = $ '1':

$$D_3 = B_3 \cdot \overline{B}_2 \cdot \overline{B}_1; \quad D_4 = B_3 \cdot (B_2 + B_1)$$

In a software implementation one approach can describe the BCD outputs using a CASE statement constructed from the binary inputs:

```
CASE (binary[3..0])  ;CASE statement
BEGIN        ;implementing look-up table
  10: bcd[4..0] = #b10000
  11: bcd[4..0] = #b10001
  12: bcd[4..0] = #b10010
  13: bcd[4..0] = #b10011
  14: bcd[4..0] = #b10100
  15: bcd[4..0] = #b10101
OTHERWISE:  BEGIN
            bcd[4]=0 ;set MSB to 0
            bcd[3..0] = binary[3..0]
            END ;and copy across binary value
END
```

which when compiled achieves:

```
BCD[4]  = binary[1] * binary[3] + binary[2] * binary[3]
BCD[3]  = binary[3] * / binary[1] * / binary[2]
/BCD[2] = / binary[2] + / binary[1] * binary[3]
BCD[1]  = binary[1] * / binary[3] + / binary[1] * binary[2] *
                            binary[3]
/BCD[0] = / binary[0]
```

This is the same result as by K-map extraction.

7.3.2.3 *Minimisation by mathematics*

An alternative method for designing the binary-to-BCD converter can be derived from examining the required functionality. If we inspect the BCD code it is the same as the binary input (though with an extra '0' as the MSB) for the first nine codes from 0000_2 up to and including 1001_2. When the binary number is ten or more then the output is the input minus ten with the extra bit set to '1'. This can be summarised in software by denoting

the four-bit binary input as input[3:0] and the five-bit BCD output as output[4:0]. The MSBs are input[3] and output[4].

```
IF(input>=0 AND input<=9) THEN BEGIN ;copy input to the output
                        output[4] = 0 ;MSB = 0
                        output[3:0] = input[3:0]
                        END
                   ELSE BEGIN ;subtract ten from input
                        output[4] = 1 ;MSB =1
                        output[3:0] = input[3:0] - 10
                        END
```

Again, this does not usually lend itself to a direct software description if the logic design tool does not include support for arithmetic. It does, however, lead to a possible hardware implementation using an adder to subtract ten from the number, a multiplexer to choose either a copy of the input or the input with ten subtracted from it, and a circuit to generate the control signal for the multiplexer (Figure 7.9).

Figure 7.9 Arithmetic circuit overview

This gave the possibility for using chips in an medium-scale integration (MSI) solution since the 74157 was a quad 2-1 multiplexer and a 74283

was a four-bit adder circuit. The control signal for the multiplexer is '1' to indicate when the input number is between ten and 15, and '0' otherwise. This is actually the signal $D_4 = (B_1 + B_2) \cdot B_3$ in the combinational design since this signal is '1' and '0' for the appropriate input ranges. This gives a control signal to the multiplexer choosing the four input bits direct when '0', and the outputs of the adder when '1'. D_4 is also the MSB of the five-bit BCD output. The MSI solution would have been rather profligate since would have needed three chips at least, whereas we have already seen that the whole design can fit within a single chip. The fact that the LSB of the BCD output is identical to the LSB of the binary input (as shown in the previous section) reflects this. A single chip confers the standard advantages of a programmable implementation. It is actually possible to reduce the adder/multiplexer circuit to equal that derived by combinational logic design. This is achieved by using complement arithmetic for minimisation, which is actually suited to algebraic minimisation of circuits which can be expressed mathematically.

The adder circuit adds -1010_2 to the binary input and can in principle be reduced since we are subtracting a constant number using a generalised adder (which works for any number). It can be minimised by using either binary subtraction or complement arithmetic. We shall use two's complement arithmetic given its practical advantages. The two's complement of a number is given by inverting each bit and then adding one. In order to subtract 10_{10} from the binary input we need to perform:

$$
\begin{array}{ccccccccccccccccc}
B_3 & B_2 & B_1 & B_0 & = & B_3 & B_2 & B_1 & B_0 & & & = & B_3 & B_2 & B_1 & B_0 \\
-1 & 0 & 1 & 0 & + & 0 & 1 & 0 & 1 & + & 1 & & +0 & 1 & 1 & 0 \\
& & & & & & & & & & & & \overline{} \\
& & & & & & & = & D_3' & D_2' & D_1' & D_0' &
\end{array}
$$

D_0' is given by the addition of B_0 plus 0. Summation is the XOR function and with one input '0' the XOR gate acts as a buffer,

$$D_0' = B_0, \text{ with a carry } = \text{'0'}.$$

D_1' is given by B_1 plus 1 with the carry from the previous stage. This carry is '0' and so the summation is $\overline{B_1}$ since the XOR gate acts as an inverter when one input is '1'.

$$D_1' = \overline{B_1}, \text{ with a carry } = B_1$$

D'_2 is given by B_2 plus 1 with the carry from the D'_1. This carry is B_1 and so the result of summation is B_1 XOR B_2.

$$D'_2 = B_1 \cdot B_2 + \overline{B}_1 \cdot \overline{B}_2, \text{ with a carry } = B_1 + B_2$$

D'_3 is given by B_3 plus 0 with the carry from the D'_2. This carry is generated when either B_2 or B_1 is '1' so this carry is $B_1 + B_2$ and so the summation is then $(B_1 + B_2)$ XOR B_3.

$$D'_3 = (B_1 + B_2) \oplus B_3, \text{ generating a carry } = (B_1 + B_2) \cdot B_3$$

We can ignore the sign bit in the result since we only use the output of the subtracter when the binary input is between 1010_2 and 1111_2 when the subtracter gives a positive output. We can then ignore any carry generated by D'_3.

This gives a minimised circuit for the adder. It can be minimised only because we are subtracting a constant, specified number. We can also minimise the whole system. The multiplexer gates through either the binary input or the output from the subtracter. The LSB is the same in both cases so one of the 2-1 multiplexers is not needed at all. The multiplexer outputs are D_3, D_2, D_1, and D_0 chosen by a control signal. The control signal $(B_1 + B_2) \cdot B_3$, which when '0' selects the binary input and when '1' selects the subtracter output.

For the LSB $\quad D_0 = \overline{\text{control}} \cdot B_0 + \text{control} \cdot D'_0 \quad$ where $D'_0 = B_0$

so $\qquad\qquad D_0 = \overline{\text{control}} \cdot B_0 + \text{control} \cdot B_0 = B_0$

For D_1 $\qquad D_1 = \overline{\text{control}} \cdot B_1 + \text{control} \cdot D'_1$

$\qquad\qquad\qquad = \overline{\text{control}} \cdot B_1 + \text{control} \cdot \overline{B}_1$

$\qquad\qquad\qquad = \overline{(B_1 + B_2) \cdot B_3} \cdot B_1 + (B_1 + B_2) \cdot B_3 \cdot \overline{B}_1$

by De Morgan's law

$\qquad\qquad\qquad = (\overline{B}_3 + \overline{B}_2 \cdot \overline{B}_1) \cdot B_1 + B_2 \cdot B_3 \cdot \overline{B}_1$

$\qquad\qquad\qquad = \overline{B}_3 \cdot B_1 + B_2 \cdot B_3 \cdot \overline{B}_1$

For D_2, $\qquad D_2 = \overline{\text{control}} \cdot B_2 + \text{control} \cdot D'_2$

$\qquad\qquad\qquad = \overline{\text{control}} \cdot B_2 + \text{control} \cdot (B_1 \cdot B_2 + \overline{B}_1 \cdot \overline{B}_2)$

$\qquad\qquad\qquad = \overline{(B_1 + B_2) \cdot B_3} \cdot B_2 + (B_1 + B_2) \cdot B_3$

$\qquad\qquad\qquad \quad \cdot (B_1 \cdot B_2 + \overline{B}_1 \cdot \overline{B}_2)$

by De Morgan's law

$$= (\overline{B}_3 + \overline{B}_2 \cdot \overline{B}_1) \cdot B_2 + B_3 \cdot B_2 \cdot B_1$$
$$= \overline{B}_3 \cdot B_2 + B_1 \cdot B_2$$

For D_3

$$D_3 = \overline{(B_1 + B_2) \cdot B_3} \cdot B_3 + (B_1 + B_2) \cdot B_3$$
$$\cdot (B_1 + B_2) \oplus B_3$$

which reduces to

$$= \overline{B}_1 \cdot \overline{B}_2 \cdot B_3$$

These reduced equations are identical to the earlier combinational design. This shows not only that there is redundancy within the MSI solution, but also how mathematics can be used to reduce circuits. Note the duality between hardware and software: all design approaches have reached the same result.

7.4 Concluding Comments and Further Reading

This chapter has introduced the main number coding systems used in computer and binary arithmetic. Coding systems find much wider use in communications systems to improve efficiency and for reasons of data security. The coding system always needs to be deciphered so that we can tell exactly what the code meant. A hardware implementation will be preferred due to speed though the complexity of the coding technique might require a software solution, perhaps even implemented using a parallel processing system to achieve a required speed. The examples included here have only concentrated on simple techniques for basic code conversion but form the basis for extension to more complex systems.

The basis of computer arithmetic has been introduced. The most versatile system is floating-point arithmetic and this can be implemented in software and/or hardware in computer systems to handle the arithmetic processes (hardware is preferred for speed). There are other programmable logic chips, which include XOR gates (instead of the OR gates and sometimes called *arithmetic* devices) that better suit the design of counters. There are also more sophisticated design techniques for arithmetic circuits though these are beyond the scope of this text.

If you would like to find more details concerning arithmetic circuits and their implementation some books now concentrate on computer arithmetic.

They are older texts, but principles don't change. Parhami (2010), Vladutiu and Du Iu (2012), and Ercegovac and Lang (2003) are all good sources. Naturally, none can divert to texts on computer architecture for which there are many since arithmetic is central to a computer's organisation. A well-established and excellent text here is Patterson and Hennessy (2008).

7.5 Questions

Solutions in Appendix 1.

Q7.1 Design a circuit to convert a three-bit natural binary number to a three-bit Gray code binary number.

Q7.2 A circuit is required which divides two two-bit binary numbers AB and CD to provide a four-bit binary result $WX.YZ$ ($AB/CD = WX.YZ$). The result is fixed point with two bits for the integer part WX and the fractional part YZ. Design a circuit to implement this function and implement it using standard combinational circuitry. Implement it again in programmable logic and show that the results are the same.

Q7.3 Design a circuit using basic binary subtraction, which subtracts ten (expressed as a binary number) from a four-bit binary number.

Q7.4 Design a circuit using one's complement arithmetic, which subtracts a four-bit binary number from ten expressed as a binary number (assume the four-bit binary number is greater than 10_{10}).

Q7.5 A circuit is required to convert single four-bit numbers (between 0 and 9) to a coded output form as:

Input number	Five-bit output				
0	0	0	0	1	1
1	0	0	1	0	0
2	0	0	1	0	1
3	0	0	1	1	0
4	0	0	1	1	1
5	0	1	0	0	0
6	0	1	1	1	1
7	1	0	0	0	1
8	1	0	0	1	1
9	1	0	1	0	1

By examining this table construct a software statement that expresses how the conversion from input to output works. Implement the circuit using:

(i) A memory chip.

(ii) An adder, a comparator and multiplexers.

(iii) The form of code used earlier.

Compare and contrast your solutions.

Q7.6 A circuit is required to add or subtract two two-bit binary numbers A and B where a mode signal M selects addition or subtraction. The signal M is LOW ($M = $ '0') to select addition, to provide $A + B$, and HIGH ($M = $ '1') to select subtraction, to provide $A - B$. The result is to be provided as a three-bit binary number (assume $A \leq B$). Design a circuit to implement this function by using:

(i) Basic binary arithmetic.

(ii) Two's complement arithmetic.

(iii) Confirm your results are the same for (i) and (ii) for one bit of the three-bit result.

Appendix 1

Abbreviated Solutions

Chapter 1

Q1.1 $5/10^2$ A

Q1.2 $V_1 \sim 5\,V$; $V_2 \sim 0\,V$

Q1.3 (c) 4.95 V; (d) 0.5 V

Chapter 2

Q2.1 These can be shown using the laws of Boolean algebra.

Q2.2 The truth table can be constructed by analysing each individual element in the function and then joining these together, using the OR function, to derive the truth table for f.

Inputs				$\overline{A \cdot C} \cdot (B \oplus D)$			$A \cdot \overline{C} \cdot \overline{(B \oplus \overline{D})}$			$B \cdot \overline{C} \cdot D$	$\overline{B + C + D}$	f
A	B	C	D	$\overline{A \cdot C}$	$B \oplus D$		$A \cdot \overline{C}$	$\overline{(B \oplus \overline{D})}$				
0	0	0	0	1	0	0	0	0	0	0	1	1
0	0	0	1	1	1	1	0	1	0	0	0	1
0	0	1	0	1	0	0	0	0	0	0	0	0
0	0	1	1	1	1	1	0	1	0	0	0	1
0	1	0	0	1	1	1	0	1	0	0	0	1
0	1	0	1	1	0	0	0	0	0	1	0	1
0	1	1	0	1	1	1	0	1	0	0	0	1
0	1	1	1	1	0	0	0	0	0	0	0	0
1	0	0	0	1	0	0	1	0	0	0	1	1
1	0	0	1	1	1	1	1	1	1	0	0	1
1	0	1	0	0	0	0	0	0	0	0	0	0
1	0	1	1	0	1	0	0	1	0	0	0	0
1	1	0	0	1	1	1	1	1	1	0	0	1
1	1	0	1	1	0	0	1	0	0	1	0	1
1	1	1	0	0	1	0	0	1	0	0	0	0
1	1	1	1	0	0	0	0	0	0	0	0	0

Q2.3 Go on, use Logisim.

Q2.4 Decoder outputs $ABC = 000, 010$ and 110 are grouped using an OR gate to provide f. No connection is made to the unused outputs.

Q2.5 Multiplexer inputs $ABC = 000, 010$ and 110 are connected to '1' and the other inputs are connected to '0'. The multiplexer output is f.

Q2.6 The solution to Q2.3 is a minimised decoder/multiplexer solution, but without the unused (and thus redundant) outputs or inputs. For the decoder, the outputs for $ABC = 010, 011, 100, 101$ and 111 are unused so the (OR) gates proving these outputs are redundant. Similarly, the terms $ABC = 010, 011, 100, 101$, and 111 in the multiplexer are unused, since their input is '0'. Further, we can remove the inputs '1', ending up with the circuit for Q2.3.

Q2.7 ```
f = ((NOT(A) AND NOT(B) AND C) OR (NOT(A) AND B AND
 NOT(C)) OR (A AND NOT(B) AND NOT(C))) OR
 (A AND B AND C))
```

Alternatively,

```
IF ((NOT(A) AND NOT(B) AND C) OR (NOT(A) AND B AND
 NOT(C)) OR (A AND NOT(B) AND NOT(C))) OR
 (A AND B AND C)) THEN f=1 ELSE f=0
```

Q2.8  ```
CASE (A,B,C)
   BEGIN
   000: f=0
   001: f=1
   010: f=1
   011: f=0
   100: f=1
   101: f=0
   110: f=0
   111: f=1
   END
```

This could be nested CASE statements, but that would be less elegant.

Q2.9 For $I =$ driver, $S =$ seat belt, $D =$ door, and $A =$ safety

$$f = I \cdot S \cdot D \cdot \overline{A}$$

Chapter 3

Q3.1 By extracting two prime implicants from the K-map the solution is:

$$f = \overline{A} \cdot B \cdot \overline{D} + A \cdot \overline{B} \cdot D$$

This is a minimum result in terms of the specified gates. To implement the function using NOR gates we can either extract the '0's or just twice invert f.

Q3.2 For a hazard-free solution, by extracting four prime implicants from the K-map, one of which is a bridging term:

$$f = A \cdot \overline{C} + A \cdot \overline{D} + \overline{A} \cdot \overline{B} \cdot D + \overline{B} \cdot \overline{C} \cdot D$$

Q3.3 By using De Morgan's law and by expanding the EXOR term, then by using a K-map, we obtain a minimal solution as:

$$f = \overline{B} \cdot \overline{C} + B \cdot C$$

Q3.4 Again by applying De Morgan's law and by expanding the EXNOR terms, then by using a K-map we obtain:

$$f = \overline{A} \cdot \overline{C} + A \cdot C$$

Q3.5 The level of logic is 3. The worst-case propagation delay is conservatively 33 ns and pedantically 32 ns.

Q3.6 A static hazard can exist between the first two terms and the last two terms, these are where bridging terms can be used. The hazard-free output f_{HF}

$$f_{HF} = \overline{A} \cdot C \cdot D + B \cdot C \cdot \overline{D} + \overline{A} \cdot B \cdot C + A \cdot \overline{C} \cdot D + A \cdot \overline{B} \cdot \overline{C}$$

The level of logic is 3 corresponding to a worst-case delay of 30 ns. The outputs can be used after this.

Chapter 4

Q4.1 One solution is to couple an inverter to the output of a two-input NAND gate.

Q4.2 Connect one input high, then the transistors connected to this input are permanently ON (NMOS) or permanently off (PMOS), and so are redundant.

Q4.3 $400\,\Omega$.

Q4.4 A parallel connection of series NMOS transistors connected to GND gives a LOW output for $f = \overline{A \cdot B \cdot C \cdot D}$ and a series pair of parallel PMOS transistors connected to V_{DD} gives the HIGH output for $f = (\overline{A} + \overline{B}) \cdot (\overline{C} + \overline{D})$

Q4.5 The music gives the age of this technology. Two diodes in parallel provide an OR gate. The output needs to be connected to ground though a resistor to give the LOW state. Note that the fan-out is <1 since the voltage dropped across the diode reduces the voltage.

Q4.6 We're up to the 1970s now. A multi-emitter transistor where the inputs are connected to the multiple emitters, the base to $+V$ (through a resistor) and the collector to an inverting transistor amplifier, gives the basis of a transistor NAND gate. The output is that of the inverter.

Chapter 5

Q5.1 Use pull-down resistors (connected to 0 V rather than $+V$) on the bistable inputs and the single pole connection on the switch to $+V$.

Q5.2 Swap the NAND for NOR and check the level of the clock which enables the circuit.

Q5.3 $\frac{1}{23} \times 10^9$ Hz. (T_H is not in the calculation and there is no combinational logic.)

Q5.4 The device should toggle when T is HIGH and latch when T is low, so the D input is formed from the fed-back bistable outputs as $D = \overline{T} \cdot Q + T \cdot \overline{Q}$. Given the two logic gates, we have $\frac{1}{10+13+24} \times 10^9$ Hz.

Q5.5 No, the D input changes within T_{SU} so the circuit will not operate correctly.

Q5.6 This needs an RC network and the pushbutton charges up the capacitor, which then discharges though the resistor until a threshold is reached. Normally, this uses a Schmitt trigger inverter.

Chapter 6

Q6.1 For a natural binary-encoded sequence of states where 111 is unused (going to 000), and for a coding $D_2 = \text{MSB}$ and $D_0 = \text{LSB}$

$D_2 = Q_2 \cdot \overline{Q}_1 + \overline{Q}_2 \cdot Q_1 \cdot Q_0$;
$D_1 = \overline{Q}_1 \cdot Q_0 + \overline{Q}_2 \cdot Q_1 \cdot \overline{Q}_0$;
$D_0 = \overline{Q}_2 \cdot \overline{Q}_0 + \overline{Q}_1 \cdot \overline{Q}_0$.

Q6.2 The T inputs are $T_2 = \text{MSB}$ and $T_0 = \text{LSB}$.

$T_2 = Q_1 \cdot Q_0 + Q_2 \cdot Q_1$;
$T_1 = Q_0 + Q_2 \cdot Q_1$;
$T_0 = \overline{Q}_2 + \overline{Q}_1$.

Q6.3 Use the ripple counter (Section 6.4.1) with \overline{CLR} asynchronous inputs connected to

$$\overline{CLR} = \overline{Q_2 \cdot Q_1 \cdot Q_0}$$

Q6.4 The T-type solution and the D-type are identical. The unused state was not clearly specified for the T-types. The (simpler) asynchronous design enters the unused state and then resets leading to potential unreliability.

Q6.5
```
CASE (Q[2..0])
   BEGIN
   #b000: BEGIN start=0
             IF (go) THEN BEGIN Q[2..0]=#b000 END
                     ELSE BEGIN Q[2..0]=#b001 END
          END
   #b001: BEGIN start=0; Q[2..0]=#b010 END
   #b010: BEGIN start=1; Q[2..0]=#b011 END
   #b011: BEGIN start=1; Q[2..0]=#b100 END
   #b100: BEGIN start=0; Q[2..0]=#b000 END
   #b101,#b110, #b111: BEGIN start=0 Q[2..0]=#b010 END
```

Q6.6
```
IF (/Q[2]*/Q[1]*/Q[0])
              THEN
              BEGIN start=0;
                    IF (go) THEN BEGIN Q[2..0]=#b000 END
                            ELSE BEGIN Q[2..0]=#b001 END
              END
ELSE
BEGIN IF (/Q[2]*/Q[1]*Q[0])
                 THEN BEGIN start=0;
                      Q[2..0]=#b010
              END
```

```
            ELSE
            BEGIN IF (/Q[2]*Q[1]*/Q[0])
                              THEN BEGIN start=1;
                                   Q[2..0]=#b011
                              END
        ELSE
        BEGIN IF (/Q[2]*Q[1]*Q[0])
                          THEN BEGIN start=1;
                               Q[2..0]=#b100
                          END
            ELSE
            BEGIN IF (Q[2]*/Q[1]*/Q[0])
                    THEN BEGIN start=0;
                         Q[2..0]=#b000
                    END
                ELSE
                BEGIN Start=0;
                      Q[2..0]=#b000
                END
            END
        END
    END
END
```

Q6.7 The asynchronous counter erroneously activates decoded outputs so the synchronous counter would appear a wise choice.

Q6.8 The overall design is functionally equivalent to multiplexer selecting an output from a set of inputs in sequence controlled by an up/down counter. Using a Modulo-6 counter and decoding the outputs ($Q_2 Q_1 Q_0 = 000$ to select A, $Q_2 Q_1 Q_0 = 001$ and 011 to select B, $Q_2 Q_1 Q_0 = 100$ and 110 to select C, $Q_2 Q_1 Q_0 = 101$ to select D) where the sequence is $Q_2 Q_1 Q_0 = 000, 001, 100, 101, 110, 011$ with states 010 and 111 unused (returning to 000) the solution is:

$D_2 = \overline{Q}_1 \cdot Q_0 + Q_2 \cdot \overline{Q}_1;$
$D_1 = Q_2 \cdot \overline{Q}_1 \cdot Q_0 + Q_2 \cdot Q_1 \cdot \overline{Q}_0;$
$D_0 = \overline{Q}_1 \cdot Q_0 + Q_2 \cdot \overline{Q}_1;$
Select $A = \overline{Q}_2 \cdot \overline{Q}_0;$
Select $B = \overline{Q}_2 \cdot Q_1;$
Select $C = Q_2 \cdot \overline{Q}_0;$
Select $D = Q_2 \cdot Q_0.$

Q6.9 For Q_0 latching the input data and Q_1 latching Q_0 (in the shift register) and for a two-bit counter which counts in natural binary we have:

$$D_1 = \overline{Q}_1 \cdot Q_0 + Q_0 \cdot S/\overline{C} + Q_1 \cdot \overline{Q}_0 \cdot \overline{S/\overline{C}};$$
$$D_0 = S/\overline{C} \cdot data + \overline{Q}_0 \cdot \overline{S/\overline{C}}.$$

The max frequency $= 21.3$ MHz

Q6.10 Use a three-bit bidirectional shift register and a Modulo-3 counter to clock it. A natural binary Modulo-3 counter is:

$$D_1 = \overline{Q}_1 \cdot Q_0;$$
$$D_0 = \overline{Q}_1 \cdot \overline{Q}_0.$$

Q6.11 The output of the AND gate will be '1' every fourth clock cycle. The function is then equivalent to that of a two-bit counter, but using an extra bistable, notwithstanding the use of T-types.

Chapter 7

Q7.1 The solution for bits D, E, and F converted from bits A,B, and C is
$F = B \oplus C; E = A \oplus B;$ and $D = A;$

Q7.2 The extraction is then, via automation,

```
Z = /D * /C * B + /D * /C * A + D * C * B * /A
Y = /D * B + /D * /C * A + D * C * /B * A
/X = /B * /A + D * /B + C * /A
W = /C * A + /D * /C * B
```

Q7.3 The result is the same as in Section 7.3.2.2.

Q7.4 The result is the same as in Section 7.3.2.2.

Q7.5 Use a truth table for the memory and a circuit implementing

```
IF (input<=5) THEN output:=input+3
   ELSE output:=2*input+3;
```

using a digital comparator to give a signal for 2-1 line multiplexers to choose either the input bits or the input bits shifted left one place (equivalent to multiplication by two), followed by an adder adding three. To implement it using a technology we need to use a CASE construct expressing the truth table.

Digital Electronics: A Primer

Q7.6 (i) Use a 2-1 line multiplexer controlled by the mode signal M to choose either the output of the adder (half adder plus full adder) or the subtracter (half subtracter plus full subtracter).

(ii) Use the mode signal to gate the complemented form of B to a set of full adders and also to set the carry input to the LSB as '1'.

(iii) B_0 is clearly the same for both circuits.

References

Alvarez, A.R. (1989). *BICMOS Technology and Applications*, Kluwer Academic Publishers, Dordrecht.

Boole, G. (1958). *An Investigation of the Laws of Thought on Which are Founded the Mathematical Theories of Logic and Probabilities*, Macmillan. London.

Brayton, R.K., Hachtel, G.D., McMullen, C.T. and Sangiovanni-Vincentelli, A.L. (1984). *Logic Minimisation Algorithms for VLSI Synthesis*, Kluwer Academic Publishers, Dordrecht.

Burch, C. (n.d.). Logisim. Available at: http://ozark.hendrix.edu/~burch/logisim. Accessed 15 April 2014.

Burch, C. (2002). Logisim: A graphical system for logic circuit design and simulation, *JERIC*, 2(1), 5–16.

Burch, C. and Ziegler, L. (2004). Science of computing science (SOCS): Resources for a breadth-first introduction, *ACM SIGCSE Bulletin*, 36(1), 437–441.

Clare, C.R. (1973). *Designing Logic Systems Using State Machines*, McGraw-Hill, New York, NY.

Dodgson, C.L. (1958). *The Game of Logic*, Dover Publications, New York, NY.

Elmasry, M.I. (1983). *Digital Bipolar Integrated Circuits*, Wiley, Oxford.

Ercegovac, M.D. and Lang, T. (2004). *Digital Arithmetic*, The Morgan Kaufmann Series in Computer Architecture and Design, Morgan Kaufmann, Burlington, MA.

Floyd, T.L. (2002). *Digital Fundamentals with VHDL*, 8th edition, Prentice Hall, Upper Saddle River, NJ.

Karnaugh, M. (1953). The map method for synthesis of combinational logic circuits, *T AIEE, Commun & Electron*, 72(1), 593–598.

Katz, R.H. and Borriello, G. (2004). *Contemporary Logic Design*, 2nd edition, Prentice Hall Upper Saddle River, NJ.

Kham, M.N., El Mubarek, H.A.W., Bonar, J.M., Ashburn, P., Ward, P., Fiore, L., Petralia, R., Alemanni, C. and Messina, A. (2006). 110GHz fT silicon bipolar transistors implemented using fluorine implantation for boron diffusion suppression, *IEEE T Electron Dev*, **53**(3), 545–552.

McCluskey, E.J. (1956). Minimization of Boolean functions, *Bell System Technical Journal*, **35**(5), 1417–1444.

McCluskey, E.J. (2001). Hat Photos. Available at: http://www-crc.stanford.edu/users/ejm/hats.html. Accessed 15 April 2014.

McGeer, P.C., Sanghavi, J.V., Brayton, R.K. and Sangiovanni-Vicentelli, A.L. (1993). ESPRESSO-SIGNATURE: A new exact minimizer for logic functions, *IEEE T VLSI Syst*, **1**(4), 432–440.

Millman, J. and Grabel A. (1988). *Microelectronics*, 2nd edition, McGraw-Hill, New York, NY.

Nixon, M.S. (1995). *Introductory Digital Design*, Macmillan, London.

Parhami, B. (1999). *Computer Arithmetic: Algorithms and Hardware Designs*, Oxford University Press, Oxford.

Parker, G.J. (2004). *Introductory Semiconductor Device Physics*, Institute of Physics Publishing, London.

Patterson, D.A. and Hennessy, J.L. (2008). *Computer Organization and Design: The Hardware/Software Interface*, 4th edition, Morgan Kaufmann, Burlington, MA.

Ross, K.A. and Wright, C.R.B. (2003). *Discrete Mathematics*, 5th edition, Prentice Hall Upper Saddle River, NJ.

Roth, C.H. and Kinney, L.L. (2009). *Fundamentals of Logic Design*, 6th edition, Cengage Learning, Stamford, CT.

Sedra, A.S. and Smith, K.C. (2010). *Microelectronic Circuits*, 6th edition, Oxford University Press, Oxford.

Spencer, R. and Ghausi, M. (2003). *Introduction to Electronic Circuit Design*, Prentice Hall, Upper Saddle River, NJ.

Tocci, R., Widmer, N. and Moss, G. (2004). *Digital Systems: Principles and Applications*, Pearson Education, Upper Saddle River, NJ.

Vahid, F. (2010). *Digital Design with RTL Design, VHDL, and Verilog*, Wiley, Oxford.

Vladutiu, M. and Du Iu, M.V.L. (2012). *Computer Arithmetic: Algorithms and Hardware Implementations*, Springer, Oxford.

Wakerly, J.F. (2005). *Digital Design: Principles and Practices*, 4th edition, Prentice Hall, Upper Saddle River, NJ.

Weste, N.H.E. and Harris, D. (2010). *CMOS VLSI Design: A Circuits and Systems Perspective*, 4th edition, Addison-Wesley, Oxford.

Zwolinski, M. (2004). *Digital System Design with VHDL*, 2nd edition, Prentice Hall, Upper Saddle River, NJ.

Zwolinski, M. (2009). *Digital System Design with SystemVerilog*, Prentice Hall, Upper Saddle River, NJ.

Index

Printed in the United States
By Bookmasters